D1544222

contemporary ceramic techniques

contem

porary

ceramic

techniques

John W. Conrad

San Diego Mesa College
San Diego, California

PRENTICE-HALL, INC., Englewood Cliffs, New Jersey 07632

Library of Congress Cataloging in Publication Data

CONRAD, JOHN W.
  Contemporary ceramic techniques.

  Bibliography: p.
  Includes index.
  1. Pottery craft. I. Title.
TT920.C64   738.1′4   77-25456
ISBN 0-13-169540-1

contemporary ceramic techniques
John W. Conrad

Copyright © 1979 by Prentice-Hall, Inc., Englewood Cliffs, New Jersey 07632

All rights reserved.
No part of this book may be reproduced
in any form or by any means
without permission in writing
from the publisher.

Printed in the United States of America

10 9 8 7 6 5 4 3 2 1

PRENTICE-HALL INTERNATIONAL, INC., *London*
PRENTICE-HALL OF AUSTRALIA PTY. LIMITED, *Sydney*
PRENTICE-HALL OF CANADA, LTD., *Toronto*
PRENTICE-HALL OF INDIA PRIVATE LIMITED, *New Delhi*
PRENTICE-HALL OF JAPAN, INC., *Tokyo*
PRENTICE-HALL OF SOUTHEAST ASIA PTE. LTD., *Singapore*
WHITEHALL BOOKS LIMITED, *Wellington, New Zealand*

TT
920
C64

3481871

contents

introduction

In the years since ceramics has come of age as a creative craft, the literature has grown in many directions: how-to-do-it books, short histories, an occasional contemporary survey, and general technical outlines. However, a need exists for a photographic overview of the latest products of contemporary ceramists coupled with detailed information on new, increasingly popular ceramic techniques such as decal, luster, silkscreen, salting, photosensitizing, low-temperature glazes, and Raku. This book combines technical explanations of contemporary ceramic techniques with a number of photographs surveying contemporary ceramics.

When I planned this book I asked several ceramists to send me photographic slides of their work. After viewing several slide collections, I felt that a lack of consistency and continuity in backdrops and in photographic quality would present problems in assembling the best and most up-to-date examples of contemporary ceramics. When my wife, Barbara, suggested that I photograph the ceramics myself, I contacted several ceramists across the United States, asking to visit them and photograph their work. On research sabbatical leave from Mesa College, I traveled 14,500 miles throughout the United States, from the Southwest desert to New England, from the Blue Ridge Mountains of the Carolinas to the Pacific Northwest. The most memorable experiences were my relationships with the ceramists I met, coupled with the opportunity to see their work firsthand. Besides their love of clay, they share a love of nature and a fellowship that bridges geographic gaps. I felt honored to be part of this friendly extended family.

I photographed the ceramic forms of 100 ceramists, a total of 660 prints, many of which are included in this book. The photographed ceramics range from the functional, such as dinnerware, tea and coffee pots, planters, vases, serving plates, punch bowls, mirrors, dry weed holders, and jars; to the nonfunctional, such as free-standing, small-sized, large-sized, fountain, relief, designs, forms, and recognizable subject sculpture. The media include Raku, earthenware, stoneware, "zero" shrinkage clay, and porcelain with surfaces of natural, salt, luster, silkscreen, decal, photosensitized, textured, and low- through high-temperature glazes. The ceramists include men, women, husband-wife teams, young adults through senior citizens, teachers, researchers, high-volume producers, limited-number producers, and one-of-a-kind producers. Thus, the photographs are a cross section of professional, creative ceramics today in the United States.

Contemporary ceramics as an individual creative and expressive art is becoming more meaningful both to the maker and to the user. Ceramics today is moving from the state of being a craft alone to that of being a highly refined art, with diverse implications for both the ceramist and society. The art of painting moved from realism, prior to the invention of the camera, to a more free expression; contemporary ceramics is making the same kind of transition.

According to Herbert Read, ceramics is the simplest and most difficult of all arts: the simplest because it is the most elemental, and the most difficult because it is the most abstract. "Pottery is pure art; it is art freed from any imitative intention. Sculpture, to which it is most nearly related, had from the first an imitative intention, and is perhaps to that extent less free for the expression of the

will to form than pottery; pottery is plastic art in its most abstract essence."[1]

In contemporary society, ceramics has three distinct and obvious kinds of uses: industrial, commercial (whether mass-produced or individually made), and aesthetic (may be mass-produced, but usually hand-made). This book is devoted to creative aesthetic ceramics. All of these fields contribute to society, but their purposes and techniques are vastly different.

The field of industrial ceramics produces specific basic tools or parts, such as rocket nose cones, bearings, grinding wheels, insulation, sewer pipes, printed circuits, and storage containers; the list expands every year. Ceramic products are also manufactured for the building construction industries, for example, floor, roof, wall, and counter tile; sinks; toilet bowls; and insulation. Again, the list proliferates as construction needs increase.

Commercial ceramics generally takes the form of functional domestic items, such as lamp bases, dinnerware, coffee and tea sets, vases, serving containers, cooking containers, planters, and ashtrays. Whether mass-produced or individually created, these items are intended essentially to be used. They may have some aesthetic value as well, however, illustrating how ceramic fields may overlap.

Functional values and aesthetic values are not mutually exclusive. Aesthetic objects include functional as well as nonfunctional forms. An example is a cup form with an opening so small that to drink out of it would be difficult. Decorative steins, plates hung on walls, coffee and tea sets for display, even ashtrays are other forms that traditionally

[1] Herbert Read, *The Meaning of Art* (Baltimore: Penguin Books, 1964), p. 33.

have a function but may be prized more for their aesthetic value. While an aesthetic object can still be functional, so too can functional objects have a high aesthetic value for ceramists. Individually made tableware, for example, is found in specialty shops, ceramists' studios, craft galleries, and art exhibits.

A more esoteric form of creativity, the ceramic sculpture, is built with the same motivation as any other expressive art form. It emphasizes surface for decoration, creative expression, new combinations of forms or textures, or new interpretations. Even in traditionally functional ceramic forms intended for show in exhibits and competitions, the function of the objects is secondary to their visual impact.

Clearly, both manufactured and handmade ceramics can be either high- or low-quality. Since the turn of the century, ceramics low in both technical and aesthetic quality have flooded the market, attracting uncritical buyers. Many novice ceramists, for instance, produce heavy, uninteresting pots that they display at weekend craft fairs and local shopping malls, and even sell in gimcrack gift shops. On the other hand, commercial potteries such as Wedgwood, Lenox, Corning, and Metlock produce sets and unique pieces that have earned them a reputation for outstanding quality. The outstanding ceramists represented in this book are but a few of the ceramists who have produced consistently high-quality work over the years.

A special class of ceramics is the commissioned piece, ranging from monogrammed tableware to individually made vases, bowls, tea and coffee sets, planters, punch bowls, and plaques. In many cases the one-of-a-kind is a limited production of that particular design and/or glaze combination, which maintains its uniqueness.

Individual ceramists are an important part of the contemporary craft scene, their work a welcome relief from the sameness of the large potteries. Creative ceramists express themselves through new combinations, innovative glazes, unusual forms, unique approaches, and varied techniques. Into the prescriptions accepted by our culture they incorporate concepts from other cultures. They invent and improve clays and glazes, combine other media with clay, fuse glass to clay, and have developed "zero" shrinkage clay. In all respects they are continuously innovative.

Creative ceramists generally conform to understood craftsmanship and take pride in their endeavors, although several influences affect the outcome. Whether the ceramist's imagination has been inhibited by adherence to the aesthetic standards ingrained by the culture is unknown. Further, his products may consciously or unconsciously reflect his estimate of what can be sold. We have technical resources more advanced than those of many other contemporary cultures; we also accept more diverse ideas than our predecessors did. Thus, the creative ceramist has a widely increased number of individual expressive choices. Without a doubt, ceramists are properly part of the art world, for they are applying themselves to creative ceramics; the work of the prominent ceramists represented in this book is inspiring evidence.

This book is intended for those who have basic knowledge, skills, and techniques of the ceramic arts and would like to expand their interests into different areas of ceramics. The information on procedures illustrated in this book provides a comprehensive understanding of each ceramic area. Many parts of the book will interest novice ceramists, but I have assumed that the reader has basic understandings and skills in the use of clay, glazes, throwing, hand building, kiln firing, and photography (for photosensitized ceramics). Several ceramic techniques described in the book require specialized equipment and materials, such as photographic supplies and enlarger for photosensitized ceramics, silkscreen materials and screen for silkscreen, unusual ceramic compounds for low-temperature glazes and for lusters, kiln and tongs for Raku, and a specialized kiln for salt-firing. Researching, testing, and understanding these ceramic techniques will give you confidence to create original products.

My appreciation extends to everyone who in some way contributed to this undertaking, for no book reaches publication without the help and support of many people. In this case, the majority are the ceramists who generously gave their time, information, and permission to photograph their work. To them, many thanks!

Although I was trained as an army photographer, I found it necessary to seek help. Jan Anderson and Bill Gordon at ColoRich Color Lab, San Diego, were especially helpful. A number of companies and individuals supplied photographic equipment as well as advice and expertise, including J. P. Corey, Administrative Manager, Ehrenreich Photo-Optical Industries; Honeywell Photographic Products; Larson Enterprises; and Kling Photo Company.[2]

When I discovered that several of the ceramists I wanted to visit had moved, Anita Chmiel, Director of Member Services, American Crafts Council, helped obtain their new addresses. I am also grateful for permission to photograph the ceramic procedures and equipment of Ron Carlson (silkscreen and decal), Diane and Steve Nelson (salt-glazing), and Marcia Lame and Randy Schneider (Raku). Weighing, mixing, and applying ceramic materials for testing is tedious, and for his patience and help in the lab I am grateful to Chris Zopada. Special thanks to Nancy Bohlander for editing, proofing, and deciphering the manuscript, and to Diane Nelson for typing it. I am indebted to Mesa College for granting the ceramic research sabbatical that enabled me to visit the ceramists. Keystone Cullet Co. (Greensburg, Pa.) and Talmach Glass Supply (La Puente, Ca.) were kind enough to supply glass cullet samples for testing.

On several occasions it was necessary to photograph ceramic work on gallery display. I appreciate the cooperation of personnel in the Fine Arts Museum, Wenger Gallery, and Quay Gallery, San Francisco; and of D. G. Keigh, Curator, Memorial Arts Gallery, Rochester, New York.

My association with those at Prentice-Hall has been extremely fortunate, and their enthusiasm for ceramics has been most welcome. I heartily thank Bud Therien, Walter

---

[2] All photographs in this book were taken by John W. Conrad unless otherwise identified. The following equipment and film were used: Mamiya RB 67 camera, 120/220 film size, 2-1/2″ by 2-3/4″ format; Mamiya Sekor lens, 127 mm, f 3.8, filters, light shade, and related equipment; Kodak Ektacolor Pro S and Tri-X film; Goosen Luna-Pro light meter; Larson Enterprises Reflectasol reflector; Smith-Victor K-2 Quartz studio flood lights; and Honeywell Auto/Strobonar 782 and Strobe-Eye strobes. Photographic processing and printing were done by ColoRich, San Diego.

Welch, and the Prentice-Hall production staff for their excellent support in handling the details of publication, including photographs, drawings, formulas, and editing.

To my family—Barbara, Bill, and Kris—I dedicate this book. Their patience, understanding, and love have been the guiding spirit behind this effort.

"Architectural Panels," Robert Glover;
stoneware, natural, oxide, black semi-gloss glaze;
each 18-1/2" H × 80" W.

"Dr. Banks Blues," Michael Frimkess;
stoneware, white glaze, overglaze decoration; 19'' H × 15'' diam.

one

low-temperature glazes

Ceramic items have served people's needs for ages. We take for granted the many ceramic products available, from tile roofs, sinks, cooking pots, and serving utensils, to electrical insulators, crucibles, rocket nose cones, insulating materials, sewage pipes, electronic components, and filters, to sculpture and dinnerware. In many ways our health and comfort still depend upon the use of clay.

Since ancient times ceramic forms have served utilitarian and aesthetic needs. Clay, shaped into all types of receptacles, enabled people to prepare, store, and carry food, water, and possessions. The earliest clay forms were made either by pressing a flattened slab of clay into a basket or other improvised mold or by roughly hollowing out lumps of clay. These early ceramics were not glazed; rather, they were decorated with colored clay or minerals.

Ancient civilizations not only used clay for fundamental needs such as food storage and serving, but they also learned to create clay items for personal adornment. One of the earliest uses of glaze to decorate has been traced to the Badarian period of ancient Egypt, about 5000 B.C. Archaeologists have found glazed ceramic beads, indicating the widespread use of bracelets, necklaces, pendants, amulets, and clothing decorations. Other glazed objects discovered include animal forms and containers shaped in molds. Soft clay was formed into pendants of geometric shapes suggesting frogs, hawks, jackals, lions, birds, or parts of these animals. Beads, often irregularly cylindrical and disc-shaped, were primitive by later standards. A soft steatite stone was used to fashion the beads, which were then covered with a turquoise-colored glaze. This glaze, a compound of siliceous paste, is known today as Egyptian paste. The color ranges from deep turquoise to pale greenish-blue. Variations of this paste were applied to vases and other objects. This was the first glazing technique to be used, and it spread rapidly.

At first only a few techniques and glaze

compounds were known; today there are many ways to apply glaze and to enrich the clay surface. Over 300 ingredients can be used to make thousands of clays, colorants, stains, feldspars, frits, enamels, glass, and glazes. The author's formula collection includes over 4,000 variations.

Interest in types of glazes produced at various firing temperatures has increased over the centuries. During the past 40 years stoneware clay and glazes have become highly popular. In the late 1960s several ceramists developed a creative approach to using low-temperature white clay and glazes. Because of stoneware's limited color range, ceramists are attracted to low-temperature glazes for their bright colors and lusters.

The formulas for low-temperature glazes in this book were obtained by exchanging with other ceramists, through research in other publications, and by experimenting. These formulas, which can be the basis for your own individual glazes, provide a cross section of low-temperature glazes. The photographs in this chapter are of ceramic work that used commercial and custom-made low-temperature glazes.

Testing the Glazes

In the low-temperature firing range many commercial glazes are available, from white through the speckled colors. Some of the speckled ones give interesting effects, with a background of one color and a speckling of complementary color(s).

Many ceramists compromise between using commercial glazes and the purist approach of using only individually made glazes. They use a commercial glaze such as satin white, translucent matt, or clear gloss as the basic glaze, and add colorants to achieve various colors and finishes. Several ceramists use this "cup-and-spoon" method effectively. Its advantages are that commercial glazes are convenient to use, it eliminates the need for an expensive inventory of raw glaze materials and a weighing balance, and the glaze can be prepared faster. However, most ceramists find it cheaper and much more rewarding to make some, if not all, of their glazes. Making individual glazes, however, has some drawbacks: an inventory of many raw materials must be bought and stored; an accurate weighing balance is essential; and the ceramist needs accuracy, patience, and a knowledge of basic glaze chemistry.

Glazing is more than simply making or purchasing a glaze, slapping it on a pot, and firing it to temperature. There are many important influences on all glazes and clay bodies, including clay composition, colorants, glaze composition, type of fuel for heat, firing temperature, and character of the firing atmosphere. Several other determinants affect the surface quality of the glaze, its translucency, its color, and its ability to fuse to the clay. All these variables necessitate testing the glaze. Only by testing can you determine whether the glaze flow is sufficient; whether the surface will be smooth enough; whether glaze application imperfections blend together; whether pinholes or bubbles are minimal; whether the glaze is chemically stable during and after firing; whether the glaze will mature at the suggested firing temperature; whether the glaze will "fit" the clay body so as to not shiver, flake, or crack it; and whether the surface of the glaze has the desired quality of matt, semi-gloss, or gloss texture.

For glaze testing use two clays: some of the clay to be used, and a white clay. Many colorants found in clay bodies will directly influence the resulting glaze color. The two finished glaze results will indicate the glaze quality on the clay being used and its "true" color on white clay, for comparison and future use.

All the testings for this book were done using white porcelain clay. Porcelain was made into an "L" form about 2″ high by $1\frac{1}{4}$″ wide by $1\frac{1}{4}$″ deep (fig. 1.1). A dark brown-black engobe was painted on one-third of the greenware form, then bisque-fired to 1600° F. The glazes were weighed out into 100-gram units, a sufficient amount of water was added to give the compound a painting consistency, and the mixed compound was painted on the test form. The glaze-covered test piece was placed in an electric kiln and fired to the recommended temperature. Then the kiln was turned off and allowed to cool at its own rate. For the reduced testings the electric kiln was fired to the recommended temperature, sawdust was placed in the kiln, the ports were sealed, and the kiln was allowed to cool at its

Figure 1.2

Glaze test samples.

own rate. Figure 1.2 shows a small sample of the hundreds of tests made to achieve the glazes listed in this book.

The tests were a uniform procedure that determined each glaze formula's (1) ability to fuse to the clay, (2) smoothness of surface, (3) surface quality, (4) degree of transparency, (5) glaze flow, (6) resulting colors, (7) accuracy of suggested firing temperature, and (8) glaze crackage. This information is recorded with each formula later in this chapter.

Test Results

Information gained from glaze testing will reveal whether the formula needs alteration and any additional testing. There are several methods of altering the glaze.

TEMPERATURE

To obtain the desired results, it may be easier to abandon a glaze formula and substitute a different one than to go through the lengthy testing procedure needed to change it. For example, if a glaze is too fluid at a certain temperature firing range, increase silica or kaolin; if the glaze is too dry, decrease silica

Figure 1.1

Porcelain for testing glazes:
(A) black engobe painted on greenware;
(B) glaze over engobe;
(C) porcelain clay.

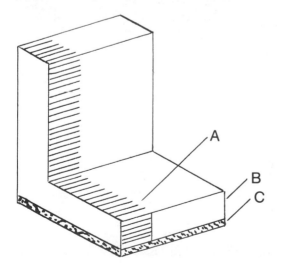

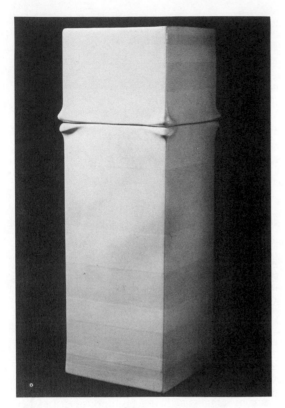

"Covered Mouth Box," Verne Funk; white earthenware, clear glaze; 17-3/4" H × 5-3/4" square.

SURFACE QUALITY

The surface quality of glazes ranges from gloss through semi-gloss, semi-matt, matt, and stony, to volcanic foam. A slight change from one surface quality to another will not require a substantial change in a glaze formula. If you increase zinc, titanium, calcium, alumina, or barium carbonate in the formula, the glaze will usually have a more matt surface. Conversely, if you remove or decrease these agents, the glaze will have a less matt surface. An underfired glaze also will often result in a matt-like surface, but the surface will often be immature with pits and craters.

GLAZE FLOW

Glaze flow or viscosity can be changed. If a glaze is too fluid, modify it by interchanging feldspars, decreasing the flux, increasing the flint or kaolin, firing at a lower temperature,

"Lelda's Hair Cut," Vincent Suez; stoneware, white glazes, overglazes; 20" square.

or kaolin. These altered compositions, however, will slightly change the characteristics of the glaze.

GLAZE MATURING RANGE

Some glaze formulas mature in a very short temperature range while others have an extraordinarily long range, as many as six cones. One fluxing compound matures at the lower temperatures and burns away at the upper ranges; the second fluxing compound acts as a refractory at the lower temperatures, while at the upper temperatures it becomes the fluxing agent. In general, most glazes have a maturing range of one to three cones.

"Envelope Box," Susanne Stephenson;
stoneware, natural, low-temperature glazes; 18-3/4" H.

applying glaze thinner, or applying a kiln wash (half kaolin and half flint) to the clay body. However, if the glaze does not have a sufficient flow the glaze application marks will show, the surface of the glaze will be irregular, and underglaze colors will not show. Improve the results by modifying the glaze, using a softer feldspar, increasing the flux, decreasing the flint or kaolin, firing at a higher temperature, using a slower firing rate, or soaking the kiln at the peak temperature.

OPACITY

To make a transparent glaze translucent, milky, opalescent, or opaque requires an opacifier such as tin, zirconium, titanium, zinc, opax, or zircopax. Even fluorspar or cryolite can be used. Various conditions cause a transparent glaze to have translucent or opaque quality, such as minute bubbles trapped within the glaze, dry matt surface, crystals forming on the surface or within the glaze, heavy concentration of metal oxides

used for colorants, and underfired glaze. When a glaze is too opaque to permit undercolors, colorants, or the clay color to show, decrease the amount of opacifier(s) in the glaze formula.

COLORS

The most common colorants used in glazes are antimony, bismuth, cadmium, cerium, chrome, cobalt, copper, iron, lead (low-temperature yellow), manganese, molybdenum, nickel, selenium, silver, sulfur (not a metal but gives amber color and is used primarily in glass), uranium, vanadium, zinc, and zirconium. Metals are the primary source of colorants used as oxides, carbonates, sulfates, and dioxides in purified form (cobalt oxide, copper carbonate, etc.) and raw or impure form (umber, yellow ochre, rutile, etc.). Used judiciously, metallic oxides and other chemical compounds can produce many pleasing varieties of low-temperature colored glazes. The results depend heavily upon the

"Deco for Kottler," Patti Warashina;
low-fire clay, glaze, luster; 36" H.

colorant(s), composition of the glaze, glaze maturing temperature, character of the firing (oxidation, neutral, or reduction atmosphere), thickness of the glaze, other glazes nearby in the kiln during the firing, and the disposition of the kiln gods. With careful measuring, blending, application of the glaze, and firing, the results should please. The blending of different colorants and opacifiers will result in a color palette of different shades, tones, and tints of colors. The strength of colorants ranges greatly from strong (cobalt oxide of 1/8 of 1 percent will give a light blue color) to weak (nickel carbonate of 3 percent will give a light gray-green color). You can obtain multicolorants from a glaze to produce various colors; for example, by using 1 percent cobalt and other colorants to produce different blues; 4 percent chrome oxide for blue-green; 2 percent nickel oxide for grayed blue; 3 percent manganese for blue-purple; 4 percent uranium for yellow-green; 5 percent red stain for red-purple; 4 percent tin oxide for light blue (sometimes pink); 3 percent yellow ochre for soft blue; 2 percent copper carbonate for turquoise; and 3 percent copper oxide for blue-green.

TOXICITY

Clay and glazes formulated with varying amounts of some ceramic compounds can be dangerous. Toxic reactions range from silicosis, caused by silica dust in the air, to cadmium poisoning. The most dangerous common materials used by ceramists are antimony, asbestos, barium, cadmium, chrome, copper, fiber glass, mica, selenium, silica, tin, and zinc. To avoid contaminating food served in finished ware, observe the following cautions for all ceramic materials, in calculating, testing, mixing, and firing.[1]

1. Glazes consist of ingredients, often including lead, that melt together and become a glass substance through high-temperature firing. Unless the glaze is properly formulated, handled, applied, and fired, it can be dangerous to both the potter and the user of the finished ware.
2. Lead, the most versatile of all the fluxes at low temperatures, has many desirable properties. However, most raw lead compounds should not be used to make glazes for food service utensils; use only the

"Obelisk," Donna Nicholas; ceramic whiteware, C/06 glaze; 28-1/2" H.

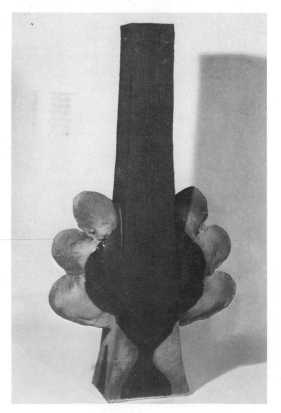

[1] See "Facts about Lead Glazes for Art Potters and Hobbyists," Lead Industries Association, Inc., 292 Madison Ave., New York, N.Y. 10017.

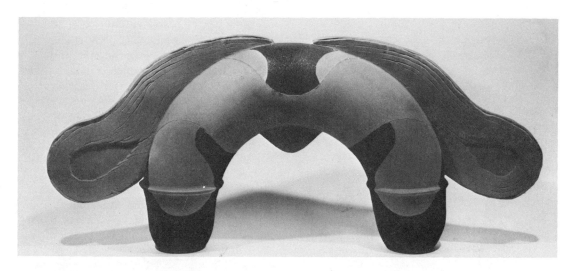

"O. M. Archetype," Donna Nicholas; ceramic whiteware, C/06 glazes; 45" W.

low-solubility-lead-type frits for glaze formulas tested for safe lead release.

3. Removing lead from a glaze will not necessarily make it safe, because antimony, barium, cadmium, chrome, copper, selenium, tin, zinc, and heavy metals are also potential poisons. Glazes are durable but not completely insoluble, and food acids will extract these potentially dangerous materials from the glaze. The amount extracted varies with the chemical composition, the ratio of ingredients, the thickness of the glaze, the temperature and time of firing, and the character of kiln atmosphere. Thorough testing will determine the solubility.

4. In some cases the mineral or metal handled may not in itself be poisonous, but its inclusion in a glaze may precipitate it or alter it so that it becomes dangerous in the fired glaze.

5. Improper balancing of the materials may result in a glaze in which some dangerous minerals or metals are not "locked in," or bonded. These unbonded materials can be leached out by mild food and drink acids, creating a potential hazard. There is no hard and fast rule that only a certain percentage of a material can be used. For example, a glaze containing high lead content and 10 percent copper oxide, fired at a very low temperature, is not safe; but 10 percent copper oxide in a stoneware glaze is generally safe.

6. There is usually little danger from lead glazes in which the lead has been fully bonded with silica. However, unbonded lead can result from underfiring or from an unbalanced lead-to-silica ratio for the particular firing temperature. Raku, overglazes, and other low-temperature raw-lead glazes are the most potentially dangerous glazes. As the temperature for firing is increased, more silica is used, giving a better bond and thus lowering the solubility of the lead.

7. All raw lead-based glazes should be tested to determine lead release. Any change in the formula or blending of "safe" lead-based glazes will change the balance of ingredients; thus, retesting is necessary. The amount of lead dissolved from the surface of a fired glaze is usually small; however, lead is cumulative in the body and will

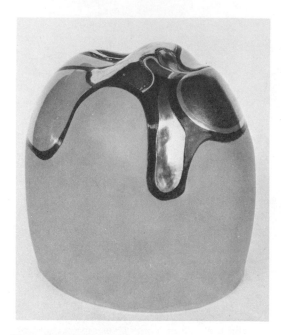

"Eagle Rock," Ralph Bacerra;
earthenware, luster, matt glaze, enamel; 10" H.

"Martha" (Great Americans Series), Victor Spinski;
low-temperature-fired stoneware,
yellow glaze, luster, decal;
17-1/2" H × 16-1/4" W.

build up to a poisonous level. Acid in foods like tomatoes, vinegar, tea, coffee, and grapefruit dissolves lead and other dangerous materials. When in doubt, do not use the glaze on food service utensils.

8. Use safety and hygiene precautions in the glazing and clay making areas. The dust from silica can cause silicosis; other dusts, such as asbestos, fiber glass, mica, and zinc, are equally dangerous. Wear a dust mask; keep dust to a minimum; and use dust-ventilating fans. Label all chemicals clearly. Forbid smoking, eating, and drinking in the glazing, clay making, and kiln rooms.

9. Indoor gas kilns should be vented and hooded; the visible flames and smoke make this obvious. Electric kilns, however, are also potentially dangerous and should at least be ventilated. Depending upon the glaze composition, antimony, barium, cadmium, carbon dioxide, carbon monoxide, chlorine, chrome, copper, lead, selenium, and zinc gases may be released from clay and glaze during firing. Working in a closed room without proper ventilation under such conditions could be extremely dangerous.

10. Before formulating glazes intended for food contact surfaces, read "FDA Laboratory Information Bulletin No. 834"[2] and "Lead Glazes for Dinnerware."[3]

More and more states are adopting the FDA test and guidelines as the standard for all lead-containing glazed ceramics intended for food contact. Such glazes will be required to meet the FDA standard of less than 7 micrograms of lead per milliliter of leaching

[2] "FDA Laboratory Information Bulletin No. 834," U.S. Food and Drug Administration, Division of Compliance Programs, Bureau of Foods (Washington, D.C.: Government Printing Office, n.d.).

[3] "Lead Glazes for Dinnerware," ILZRO Manual, Ceramics #1 (New York: International Lead Zinc Research Organization, Inc., 1970).

solution as determined by the FDA analysis method.

Chemistry

The purposes of a glaze are to give clay a protective surface and to enhance its appearance—both are important. A glaze is a compound of several elements that fuse into a glass-like substance. There is scarcely a glaze that contains fewer than three elements; therefore, a simple explanation of the entire field of glaze compositions is not possible. It is interesting to note that many established glazing procedures could not for a long time be explained scientifically. Only in the last two decades has the scientific community rapidly caught up with empirical knowledge. Many of the latest inventions and developments in the art of glazes are the result of advanced scientific approaches.

At one time ceramists thought that knowledge of the chemistry and properties of con-

stituent elements would suffice to predict and explain the resulting properties of the glaze(s). We now know that we still lack important information that would explain how the atoms of the elements in the glaze compound are interlinked. Such information would reveal exactly how one element influences another.

Most glaze formulas invented today are the result of trial and error. A ceramist refines a formula, using local chemical compounds

"Only Bastards Play War," Vincent Suez; stoneware, white glaze, overglazes; 38" H.

"Grannie's Pot," Tom Manhart; low-fire talc body, commercial glazes, luster, decal; 4" H × 6" W.

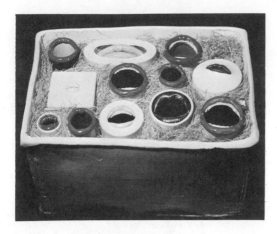

"Box of Mouth Pots," Verne Funk;
white clay, low-fire glazes, lusters;
9-1/2" H × 15-1/2" × 11".

and his own glazing and firing techniques, until it meets his needs on the particular clay body. When another ceramist uses the same formula, a new set of variables arises. Glazing

"Cookie Plate," Verne Funk;
white clay, clear low-fire glaze over pencil;
15" diam.

techniques, glaze thickness, substitution of chemical compounds, clay body, and firing all affect the outcome. No two ceramists use exactly the same materials and procedures.

Any glaze formula, regardless of its source, should be tested using a standard clay body, glaze application technique, glaze thickness, and kiln firing. The glaze can be adjusted or augmented to obtain certain results. Numerous variations can be obtained from just one glaze, for example, by (1) introducing or removing opacifiers, (2) substituting other fluxes to increase or decrease the glaze flow, (3) introducing or removing matting agents, (4) using colorants in various combinations and percentages, (5) using stains on the ceramic form or over the glaze to alter glaze color and texture, or (6) varying the kiln firing and character of the kiln atmosphere. Such changes can produce 50 or more variants of a single glaze.

Figure 1.3 is a chart of the glaze formulas. The formulas are organized into the temperature groups of Cone/013 and under, Cone/012, Cone/011, Cone/010, Cone/09, Cone/08, Cone/07, Cone/06, Conc/05, and Cone/04; subdivided as to gloss, semi-gloss, semi-matt, and matt surfaces; further subdivided as to transparent, translucent, and opaque qualities; and then distinguished by color: colorless, white, yellow, orange, red, purple, blue, green, brown, and black.

Figure 1.3 (*pages 11–41*)

Glaze Formulas. Explanation: *Temperature* refers to the temperature range recommended for firing. *Surface* is the surface quality of gloss, semi-gloss, semi-matt, matt, or stone at the temperature fired for the test. *Fluidity* refers to the fluid quality of the glaze at the tested temperature. *Stains show* refers to the ability of dark understains to show through the glaze. *Opacity* refers to the degree of opacity of the glaze, whether opaque, translucent, or transparent. *Color* is the resulting color, obtained through oxidation and/or reduction firing.

Cone/013 and Lower

G 1001 Milky White

Frit #25		Temperature:	C/016–014
(Pemco)	82.2	Surface	
Lithium		@ C/016:	Gloss
carbonate	9.0	Fluidity:	Some
Whiting	4.9	Stains show:	Yes
Kaolin	3.9	Opacity:	Translucent
	100.0	Color	
		/oxidation:	Milky
		/reduction:	Milky
		Note:	Cracks

G 1002 Clear

Frit #25		Temperature:	C/014–012
(Pemco)	30	Surface	
Lithium		@ C/014:	Semi-gloss
carbonate	30	Fluidity:	Some
Colemanite	30	Stains show:	Yes
Flint	10	Opacity:	Transparent
	100	Color	
		/oxidation:	Clear
		/reduction:	Clear
		Note:	Cracks

G 1003 Semi-Gloss

Flint	28	Temperature:	C/014–010
Calcined borax	28	Surface	
Soda ash	18	@ C/014:	Semi-gloss
Potassium		Fluidity:	Some
carbonate	8	Stains show:	Yes
Zinc oxide	8	Opacity:	Transparent
Whiting	8	Color	
Boric acid	2	/oxidation:	Clear
	100	/reduction:	Clear
		Note:	Cracks

G 1004 White

Frit #14		Temperature:	C/015–08
(Hommel)	70	Surface	
Sodium		@ C/015:	Satin matt
silicate (dry)	15	Fluidity:	Little
Flint	10	Stains show:	Darks show
Kaolin	5	Opacity:	Opaque
	100	Color	
		/oxidation:	White
		/reduction:	White
		Note:	Cracks; apply even coat

G 1005 True Yellow

Frit #25	60.1	Temperature:	C/015–013
Lithium		Surface	
carbonate	16.0	@ C/015:	Gloss
Colemanite	14.9	Fluidity:	Some
Yellow stain	7.0	Stains show:	Yes
Ball clay	2.0	Opacity:	Transparent
	100.0	Color	
		/oxidation:	Yellow
		/reduction:	True yellow
		Note:	Cracks

G 1006 Orange-Red

White lead	90	Temperature:	C/013–012
Kaolin	5	Surface	Gloss/semi-
Chromium	3	@ C/013:	gloss
Tin oxide	2	Fluidity:	Fluid
	100	Stains show:	Darks show
		Opacity:	Opaque
		Color	
		/oxidation:	Bright orange-red
		/reduction:	Do not reduce

"Ceramic Form," Ka-Kwong Hui; earthenware, overglaze, luster; 19-1/4" H × 18" W.

G 1007 Slight Blue

Calcined			Temperature:	C/018–014
borax	27.0		Surface	Gloss/semi-
Flint	27.0		@ C/018:	gloss
Soda ash	13.0		Fluidity:	Little
Potassium			Stains show:	Yes
carbonate	10.9		Opacity:	Transparent
Whiting	7.9		Color	
Colemanite	7.9		/oxidation:	Slight blue
Sulfur,			Note:	Cracks
flowers	3.0			
Sodium				
sulfate	3.0			
Copper				
carbonate	.3			
	100.0			

G 1008 Medium Blue

Colemanite	40		Temperature:	C/018–015
Frit #25			Surface	
(Pemco)	25		@ C/018:	Semi-matt
Cryolite	10		Fluidity:	Little
Lithium			Stains show:	No
carbonate	10		Opacity:	Opaque
Flint	10		Color	
Zircopax	3		/oxidation:	Medium
Copper				blue
carbonate	2		/reduction:	Medium
	100			blue

G 1009 Light Grass Green

White lead	80		Temperature:	C/014
Kaolin	10		Surface	
Tin oxide	5		@ C/014:	Gloss
Whiting	4		Fluidity:	Fluid
Copper			Stains show:	Yes
carbonate	1		Opacity:	Transparent
	100		Color	
			/oxidation:	Light grass
				green
			/reduction:	Do not
				reduce
			Note:	Apply thin
				coat

G 1010 Medium Green

White lead	57.6		Temperature:	C/013–09
Frit #25	19.2		Surface	
Silica	16.3		@ C/09:	Gloss
Copper			Fluidity:	Very fluid
carbonate	3.9		Stains show:	Yes
Red iron			Opacity:	Transparent
oxide	2.4		Color	
Cobalt oxide	.6		/oxidation:	Medium
	100.0			green
			/reduction:	Green-black

G 1011 Yellow-Green

Frit #25			Temperature:	C/013
(Pemco)	40.00		Surface	
Colemanite	20.00		@ C/013:	Gloss
Flint	10.00		Fluidity:	Some
Kaolin	10.00		Stains show:	No
Lithium			Opacity:	Opaque
carbonate	10.00		Color	
Uranium			/oxidation:	Yellow-
yellow	8.00			green
Lead			/reduction:	Yellow-
chromate	2.00			green
Cobalt				
carbonate	.03			
	100.03			

G 1012 Greenish Yellow

Borax	30		Temperature:	C/013–010
Soda ash	20		Surface	
Kaolin	15		@ C/010:	Gloss =
Colemanite	11			thick;
Boric acid	10			semi-
Whiting	10			matt =
Lead				thick
chromate	4		Fluidity:	Fluid
	100		Stains show:	Yes
			Opacity:	Transparent
			Color	
			/oxidation:	Yellow-
				green
			/reduction:	Light
				yellow-
				green
			Note:	Cracks

"Clouds" (covered jar), Ralph Bacerra;
stoneware, overglaze, slip decoration, clear glaze;
7" H × 15-1/2" diam.

G 1013 Clear Leadless

Colemanite	60	Temperature:	C/012–010
Nepheline		Surface	
syenite	30	@ C/010:	Gloss
Kaolin	5	Fluidity:	Fluid
Flint	5	Stains show:	Yes
	100	Opacity:	Transparent
		Color	
		/oxidation:	Clear
		/reduction:	Clear
		Note:	Cracks

G 1014 Water Clear

Frit #25		Temperature:	C/012–09
(Pemco)	54.6	Surface	
Colemanite	36.6	@ C/012:	Gloss
Zinc oxide	4.9	Fluidity:	Little
Kaolin	3.9	Stains show:	Yes
	100.0	Opacity:	Transparent
		Color	
		/oxidation:	Water clear
		/reduction:	Water clear
		Noto:	Cracks

G 1015 Eggshell White

Soda feldspar	30	Temperature:	C/012–010
Calcined		Surface	
borax	30	@ C/012:	Semi-gloss
Boric acid	10	Fluidity:	None
Kaolin	10	Stains show:	Yes
Selenium	8	Opacity:	Translucent
Colemanite	5	Color	
Barium		/oxidation:	Eggshell white
carbonate	5	/reduction:	Cool white
Tin oxide	2	Note:	Some cracks
	100		

G 1016 Soft White

Flint	30.2	Temperature:	C/012–010
Lithium		Surface	
carbonate	30.1	@ C/010:	Semi-gloss
Colemanite	15.2	Fluidity:	Fluid
Whiting	9.2	Stains show:	Darks
Kaolin	8.1	Opacity:	Opaque
Bentonite	2.0	Color	
Sulphur	2.0	/oxidation:	Soft white
Antimony		/reduction:	Soft white
oxide	3.0	Note:	Cracks
Silicon			
carbonate	.2		
	100.0		

"Covered Dish for California Dinner of 80 People,"
Viola Frey;
stoneware, china paints, white matt glaze;
13-1/2" H.

G 1017 Curdle
(Long Beach State, Long Beach, Ca.)

Borax	52.3	Temperature·	C/012–04
Tin oxide	17.4	Surface	Broken matt
Soda feldspar	14.5	@ C/04:	and gloss
Kaolin	8.8	Fluidity:	Little
Zinc oxide	7.0	Stains show:	No
	100.0	Opacity:	Opaque
		Color	
		/oxidation:	White curdle
		/reduction:	Light gray curdle
		Note:	Best over dark clay; more tin, more curdle

G 1018 Pale Yellow

Red lead	61	Temperature:	C/012–010
White lead	15	Surface	
Flint	10	@ C/010:	Gloss
Kaolin	4	Fluidity:	Fluid
Boric acid	4	Stains show:	Yes
Cadmium		Opacity:	Transparent
sulfide	4	Color	
Tin oxide	2	/oxidation:	Pale yellow
	100	/reduction:	Do not reduce
		Note:	Cracks

G 1019 Yellow

Borax	36.0	Temperature:	C/012
Soda ash	18.0	Surface	
Soda feldspar	9.0	@ C/012:	Gloss
Kaolin	8.8	Fluidity:	Fluid
Boric acid	8.7	Stains show:	Darks
Potassium		Opacity:	Translucent
carbonate	8.5	Color	
Yellow stain	6.0	/oxidation:	Thick =
Opax	3.0		yellow;
Crocus martis	2.0		thin =
	100.0		rich burnt
			orange
		/reduction:	black

G 1020 Medium Yellow

White lead	27.0	Temperature:	C/012–010
Litharge	20.3	Surface	
Flint	16.9	@ C/012:	Gloss
Cornwall		Fluidity:	Some
stone	15.2	Stains show:	Darks
Boric acid	13.5	Opacity:	Translucent
Yellow stain	4.1	Color	
Tin oxide	2.0	/oxidation:	Medium
Yellow ochre	1.0		yellow
	100.0	/reduction:	Grayed
			yellow

G 1021 Naples Yellow

Lead		Temperature:	C/012–010
bisilicate	60	Surface	
Lithium		@ C/012:	Gloss
carbonate	19	Fluidity:	Some
Flint	7	Stains show:	No
Tin oxide	6	Opacity:	Opaque
Kaolin	5	Color	
Antimony		/oxidation:	Naples
oxide	3		yellow
	100	/reduction:	Grayed
			Naples
			yellow
		Note:	Cracks

G 1022 Yellow and Tan

Red lead	57	Temperature:	C/012–09
Flint	30	Surface	
Tin oxide	8	@ C/012:	Satin matt
Vanadium	5	Fluidity:	Little
	100	Stains show:	Darks
		Opacity:	Translucent
		Color	
		/oxidation:	Yellow with
			broken
			tan
		/reduction:	Brown

G 1023 Light Orange

White lead	44.4	Temperature:	C/012–010
Cullet	44.4	Surface	
Uranium	5.1	@ C/010:	Gloss
Tin oxide	3.1	Fluidity:	Little
Red lead	3.0	Stains show:	Yes
	100.0	Opacity:	Transparent
		Color	
		/oxidation:	Light
			orange
		/reduction:	Brown
		Note:	Cracks;
			grind
			mixture

G 1024 Broken Yellow-Orange

Borax	20	Temperature:	C/012–010
Soda ash	20	Surface	Gloss and
Potassium		@ C/012:	matt
carbonate	20	Fluidity:	None
Red lead	15	Stains show:	None
Kaolin	10	Opacity:	Opaque
Flint	5	Color	
Tin oxide	5	/oxidation:	Broken
Chrome oxide	5		yellow-
	100		orange
		/reduction:	Dark green

G 1025 Dark Purple

White lead	46	Temperature:	C/012–010
Flint	27	Surface	
Calcined		@ C/012:	Semi-gloss
borax	11	Fluidity:	None
Kaolin	9	Stains show:	Darks
Tin oxide	4	Opacity:	Opaque
Manganese		Color	
dioxide	3	/oxidation:	Dark purple
	100	/reduction:	Cool dark
			purple

G 1026 Dark Lavender

Calcined borax	28.0	Temperature: C/012–010
Soda ash	25.0	Surface
Lithium carbonate	15.0	@ C/010: Gloss
		Fluidity: Fluid
Potassium carbonate	15.0	Stains show: Yes
		Opacity: Translucent
Kaolin	11.0	Color
Tin oxide	3.0	/oxidation: Dark lavender
Manganese dioxide	2.7	/reduction: Brown
Cobalt carbonate	.3	Note: Cracks
	100.0	

G 1027 Turquoise Blue

Borax	46	Temperature: C/012–010
Soda feldspar	37	Surface
Calcium carbonate	8	@ C/012: Gloss
		Fluidity: Fluid
Arsenic oxide	4	Stains show: Yes
Copper carbonate	3	Opacity: Transparent
		Color
Plastic vetrox	2	/oxidation: Medium turquoise blue
	100	/reduction: Do not reduce
		Note: Cracks

G 1028 Pale Turquoise

Borax	37.0	Temperature: C/012–010
Soda feldspar	19.0	Surface
Soda ash	18.0	@ C/012: Gloss
Potassium carbonate	9.0	Fluidity: Little
		Stains show: Yes
Kaolin	8.0	Opacity: Translucent
Boric acid	7.3	Color
Tin oxide	1.0	/oxidation: Pale turquoise
Copper carbonate	.6	/reduction: Broken oxblood red
Silicon carbonate	.1	Note: Bubbles where thick
	100.0	

G 1029 Avocado Green

White lead	79	Temperature: C/012–010
Kaolin	10	Surface
Zircopax	6	@ C/012: Gloss
Nickel oxide	3	Fluidity: Fluid
Whiting	2	Stains show: Yes
	100	Opacity: Translucent
		Color
		/oxidation: Broken light avocado green
		/reduction: Do not reduce

G 1030 Light Blue-Green

Red lead	47.9	Temperature: C/012
Cullet	19.6	Surface
Cornwall stone	19.2	@ C/012: Gloss
		Fluidity: Little
Flint	9.6	Stains show: Yes
China clay	2.4	Opacity: Transparent
Copper carbonate	1.2	Color
		/oxidation: Light blue-green
Cobalt carbonate	.1	/reduction: Brown
	100.0	Note: Cracks

G 1031 Dark Green

Borax	40.0	Temperature: C/012
Soda ash	20.0	Surface
Potassium carbonate	11.5	@ C/012: Gloss
		Fluidity: Fluid
Soda feldspar	10.0	Stains show: Darks
Kaolin	10.0	Opacity: Opaque
Red iron oxide	8.0	Color
		/oxidation: Dark green
Cobalt oxide	.5	/reduction: Green-black
	100.0	

G 1032 Olive Black

White lead	51.3	Temperature: C/012
Flint	17.0	Surface
Cornwall stone	15.7	@ C/012: Gloss
		Fluidity: Fluid
Borax	10.5	Stains show: No
Zircopax	3.3	Opacity: Opaque
Red iron oxide	1.3	Color
		/oxidation: Olive black
Chrome oxide	.6	/reduction: Do not reduce
Cobalt carbonate	.3	
	100.0	

"Fish and Cup," Richard Shaw;
porcelain, decal, stains, glazes; 5-3/4" H.

G 1033 Red-Orange
(Long Beach State, Long Beach, Ca.)

White lead	81.1	Temperature:	C/011
Flint	10.1	Surface	
China clay	5.0	@ C/011:	Gloss
Chrome oxide		Fluidity:	Fluid
green	2.5	Stains show:	None
Tin oxide	1.3	Opacity:	Opaque
	100.0	Color	
		/oxidation:	Orange-red
		/reduction:	Do not reduce

G 1034 Nasturtium #3
(Long Beach State, Long Beach, Ca.)

White lead	77.2	Temperature:	C/011
Flint	10.8	Surface	
China clay	6.0	@ C/011:	Semi-matt
Chrome oxide		Fluidity:	Little
green	2.4	Stains show:	None
Tin oxide	1.8	Opacity:	Opaque
Barium		Color	
carbonate	1.8	/oxidation:	Nasturtium orange
	100.0	/reduction:	Do not reduce

G 1035 Coral Red

Red lead	62.6	Temperature:	C/011
Lead		Surface	
chromate	13.2	@ C/011:	Gloss
China clay	10.9	Fluidity:	Fluid
Flint	9.4	Stains show:	Darks
Boric acid	3.9	Opacity:	Opaque
	100.0	Color	
		/oxidation:	Coral red; green where thin
		/reduction:	Do not reduce

"Dish on a Fish," Patti Warashina; low-fire clay, underglaze, glaze; 23" H.

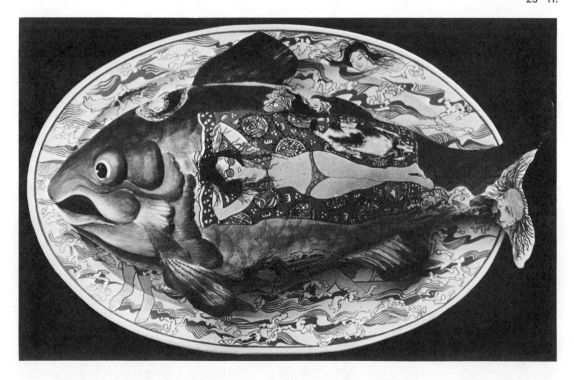

G 1036 Clear

Colemanite	60	Temperature:	C/010
Nepheline		Surface	
syenite	40	@ C/010:	Gloss
	100	Fluidity:	Little
		Stains show:	Yes
		Opacity:	Transparent
		Color	
		/oxidation:	Clear
		/reduction:	Clear
		Note:	Cracks

G 1037 Leadless Clear

Borax	50	Temperature:	C/010–09
Soda feldspar	40	Surface	
Calcium		@ C/010:	Gloss
carbonate	10	Fluidity:	Little
	100	Stains show:	Yes
		Opacity:	Transparent
		Color	
		/oxidation:	Clear
		/reduction:	Clear
		Note:	Do not
			apply
			thin

G 1038 Clear Gloss

Kona feldspar	35	Temperature:	C/010–08
Colemanite	35	Surface	
Barium		@ C/010:	Gloss
carbonate	10	Fluidity:	Some
Flint	10	Stains show:	Yes
Lithium		Opacity:	Transparent
carbonate	10	Color	
	100	/oxidation:	Clear
		/reduction:	Clear
		Note:	Some
			cracks

G 1039 Whitish

Frit #25		Temperature:	C/010
(Pemco)	60	Surface	
Whiting	15	@ C/010:	Semi-gloss
Nepheline		Fluidity:	Some
syenite	15	Stains show:	Yes
Barium		Opacity:	Translucent
carbonate	6	Color	
Zinc oxide	4	/oxidation:	Whitish
	100	/reduction:	Whitish
		Note:	Cracks

G 1040 Beautiful Yellow
(Long Beach State, Long Beach, Ca.)

White lead	67.9	Temperature:	C/010–09
Cornwall		Surface	
stone	11.9	@ C/010:	Gloss
Flint	11.9	Fluidity:	Some
Tin oxide	6.4	Stains show:	Yes
Lead		Opacity:	Transparent
chromate	1.9	Color	
	100.0	/oxidation:	Beautiful
			yellow
		/reduction:	Green
		Note:	Cracks;
			best on
			white
			clay

G 1041 Bright Yellow

White lead	67	Temperature:	C/010
Silica	29	Surface	
Kaolin	4	@ C/010:	Gloss
	100	Fluidity:	Some
		Stains show:	Yes
		Opacity:	Transparent
		Color	
		/oxidation:	Bright
			yellow
		/reduction:	Do not
			reduce
		Note:	Cracks

G 1042 Transparent Yellow

Colemanite	51	Temperature:	C/010–09
Nepheline		Surface	
syenite	28	@ C/010:	Gloss
Boric acid	10	Fluidity:	Some
Plastic vetrox	6	Stains show:	Yes
Cadmium		Opacity:	Transparent
sulfide	3	Color	
Tin oxide	2	/oxidation:	Light true
	100		yellow
		Note:	Cracks

G 1043 Naples Yellow

Lead		Temperature:	C/010–09
bisilicate	75.3	Surface	
Flint	7.6	@ C/010:	Gloss
Tin oxide	5.4	Fluidity:	Little
Kaolin	5.2	Stains show:	Darks
Lithium		Opacity:	Opaque
carbonate	4.2	Color	
Antimony		/oxidation:	Naples
oxide	2.3		yellow
	100.0	/reduction:	Naples
			yellow

G 1044 Lead Yellow

Lead oxide	46
Flint	35
Calcined borax	11
Antimony oxide	6
Kaolin	2
	100

Temperature: C/010–08
Surface
  @ C/010: Semi-gloss
Fluidity: Little
Stains show: Darks
Opacity: Translucent
Color
  /oxidation: Dark yellow
  /reduction: Dark yellow
Note: Bubbles where thick

G 1045 Opaque Yellow

White lead	51
Soda feldspar	25
Calcined clay	10
Whiting	6
Kaolin	3
Zircopax	3
Flint	2
	100

Temperature: C/010–08
Surface
  @ C/010: Semi-gloss
Fluidity: Little
Stains show: Darks
Opacity: Opaque
Color
  /oxidation: Light yellow
  /reduction: Light yellow
Note: Cracks

G 1046 Bright Orange
(Long Beach State, Long Beach, Ca.)

White lead	62.9
Flint	14.6
Lead chromate	8.8
China clay	5.4
Tin oxide	4.9
Boric acid	3.4
	100.0

Temperature: C/010–09
Surface
  @ C/010: Gloss
Fluidity: Fluid
Stains show: Yes
Opacity: Transparent
Color
  /oxidation: Bright orange
  /reduction: Do not reduce
Note: Use on white clay

G 1047 Uranium Orange

White lead	50
Flint	20
Soda ash	20
Bone ash	5
Uranium oxide	4
Tin oxide	1
	100

Temperature: C/010–08
Surface
  @ C/08: Gloss
Fluidity: Fluid
Stains show: Yes
Opacity: Transparent
Color
  /oxidation: Orange, where thick
  /reduction: Do not reduce
Note: Cracks

G 1048 Truly Orange

Frit #25 (Pemco)	56
Lithium carbonate	19
Uranium, yellow	18
Magnesium carbonate	4
Tin oxide	2
Bentonite	1
	100

Temperature: C/010
Surface
  @ C/010: Semi-gloss
Fluidity: Some
Stains show: Darks
Opacity: Translucent and opaque areas
Color
  /oxidation: Truly orange
  /reduction: Grayed truly orange
Note: Cracks

G 1049 Medium Orange

White lead	70
Flint	15
Kaolin	8
Antimony oxide	4
Chrome oxide	2
Tin oxide	1
	100

Temperature: C/010–05
Surface
  @ C/05: Semi-gloss
Fluidity: Little
Stains show: No
Opacity: Opaque
Color
  /oxidation: C/010 = medium orange; C/05 = greens
  /reduction: C/010 = medium orange; C/05 = greens

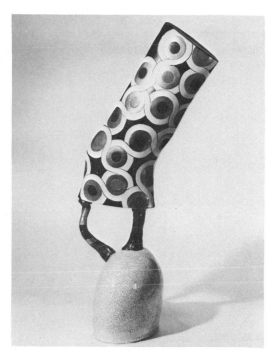

"Circle I," Ralph Bacerra;
earthenware, lusters, enamel; 22-1/2" H.

G 1062 Green

Colemanite	45.6	Temperature:	C/010–08
Plastic vetrox	19.6	Surface	
Albany slip	19.6	@ C/010:	Semi-gloss
Zinc oxide	6.5	Fluidity:	Fluid
Borax	6.5	Stains show:	Darks
Copper		Opacity:	Translucent
carbonate	2.2	Color	
	100.0	/oxidation:	Green
		/reduction:	Warm green
		Note:	Cracks

G 1063 Jade Green

White lead	65	Temperature:	C/010–09
Flint	10	Surface	
Soda		@ C/010:	Semi-matt
feldspar	10	Fluidity:	Little
Whiting	6	Stains show:	No
Rutile	6	Opacity:	Opaque
Kaolin	2	Color	
Copper oxide		/oxidation:	Jade green
black	1	/reduction:	Broken jade green
	100		

G 1064 Great Yellow-Green

Lithium		Temperature:	C/010–08
carbonate	25	Surface	
Borax	25	@ C/010:	Semi-gloss
Soda ash	25	Fluidity:	Little
Kaolin	11	Stains show:	No
Lead		Opacity:	Opaque
chromate	5	Color	
Tin oxide	5	/oxidation:	Yellow-green
Flint	4	/reduction:	Yellow-green
	100	Note:	Grind mixture

G 1065 Fluid Amber

White lead	90	Temperature:	C/010
Red iron		Surface	
oxide	6	@ C/010:	Gloss
Flint	4	Fluidity:	Fluid
	100	Stains show:	Yes
		Opacity:	Transparent
		Color	
		/oxidation:	Bright amber orange
		/reduction:	Dark amber brown
		Note:	Cracks

G 1066 Broken Tans

White lead	83	Temperature:	C/010
Kaolin	10	Surface	Broken
Red iron		@ C/010:	gloss and matt
oxide	3		
Titanium		Fluidity:	Some
dioxide	2	Stains show:	Darks
Tin oxide	2	Opacity:	Opaque
	100	Color	
		/oxidation:	Broken tan, gold, and yellow
		/reduction:	Do not reduce

G 1067 Volcanic Ash

Volcanic ash	29.5	Temperature:	C/010–08
Colemanite	19.6	Surface	
White lead	18.6	@ C/010:	Matt
Kaolin	15.7	Fluidity:	Little
Tin oxide	5.9	Stains show:	Darks
Borax	3.9	Opacity:	Opaque
Whiting	3.9	Color	
Zinc oxide	2.9	/oxidation:	Light
	100.0		broken
			gray
		/reduction:	Do not
			reduce

G 1068 Oregon Red Grain
(Long Beach State, Long Beach, Ca.)

White lead	56.6	Temperature:	C/010–04
Flint	17.4	Surface	
Cornwall		@ C/010:	Semi-gloss
stone	13.0	@ C/04:	Gloss
Black iron		Fluidity:	Little
oxide	4.8	Stains show:	Yes
Frit #25		Opacity:	Transparent
(Pemco)	4.5	Color:	C/010
Tin oxide	3.7		Brown
	100.0		streaked
			with
			orange
			C/08
			Red-
			brown
			with
			black
			C/07
			Streaked
			red, grain
			finish
			C/05
			Reduced
			= golden
			brown
			C/04
			Red and
			golden
			brown

G 1069 Clear

Colemanite	85.7	Temperature:	C/09
Whiting	8.2	Surface	
Kaolin	6.1	@ C/09:	Gloss
	100.0	Fluidity:	Little
		Stains show:	Yes
		Opacity:	Transparent
		Color	
		/oxidation:	Clear
		/reduction:	Clear
		Note:	Cracks;
			good for
			Raku

G 1070 Leadless Clear

Boric acid	61.5	Temperature:	C/09–07
Soda ash	15.4	Surface	
Flint	15.4	@ C/07:	Gloss
Magnesium		Fluidity:	Fluid
carbonate	7.7	Stains show:	Yes
	100.0	Opacity:	Transparent
		Color	
		/oxidation:	Clear
		/reduction:	Clear
		Note:	Cracks

"Ceramic Form," Ka-Kwong Hui;
ceramic whiteware, low-fire overglaze, luster;
20-1/2" H × 14" W.

G 1071 Burnt Orange

Red lead	66		
Silica	18	Temperature:	C/09
Kaolin	8	Surface	
Selenium	3	@ C/09:	Gloss
Soda ash	2	Fluidity:	Some
Cadmium	2	Stains show:	Yes
Chrome oxide	1	Opacity:	Transparent
	100	Color	
		/oxidation:	Burnt orange
		/reduction:	Do not reduce
		Note:	Cracks; best on white clay

G 1072 Orange-Red (Lukens #3)

White lead	77.8		
Flint	7.9	Temperature:	C/09
Edgar plastic		Surface	
kaolin	6.4	@ C/09:	Semi-gloss
Chrome oxide	5.5	Fluidity:	Little
Tin oxide	2.4	Stains show:	No
	100.0	Opacity:	Opaque
		Color	
		/oxidation:	Bright orange-red
		/reduction:	Do not reduce

G 1073 Medium Blue

Calcined			
borax	33	Temperature:	C/09
Flint	25	Surface	
Soda ash	25	@ C/09:	Gloss
Whiting	10	Fluidity:	Some
Zinc oxide	4	Stains show:	Yes
Red iron		Opacity:	Transparent
oxide	2	Color	
Cobalt		/oxidation:	Medium blue
oxide	1	/reduction:	Medium blue
	100	Note:	Fine cracks

G 1074 Abalone Green

Colemanite	58		
Nepheline		Temperature:	C/09–08
syenite	39	Surface	
Copper		@ C/09:	Semi-gloss
carbonate	2	Fluidity:	Little
Red iron		Stains show:	Darks
oxide	1	Opacity:	Translucent
	100	Color	
		/oxidation:	Abalone green

G 1075 Yellow-Green

Calcined			
borax	30	Temperature:	C/09
Soda ash	30	Surface	
Whiting	20	@ C/09:	Gloss
Ball clay	10	Fluidity:	Little
Tin oxide	7	Stains show:	Yes
Chrome oxide	3	Opacity:	Translucent
	100	Color	
		/oxidation:	Yellow-green
		/reduction:	Do not reduce

G 1076 Chocolate

Colemanite	40		
Nepheline		Temperature:	C/09–08
syenite	24	Surface	
Albany slip	24	@ C/09:	Semi-matt
Red iron		Fluidity:	None
oxide	12	Stains show:	No
	100	Opacity:	Opaque
		Color	
		/oxidation:	Chocolate
		/reduction:	Chocolate

G 1077 Dark Brown

Borax	39.0		
Flint	25.3	Temperature:	C/09–08
Colemanite	25.2	Surface	
Red iron		@ C/09:	Semi-gloss
oxide	10.1	Fluidity:	Little
Cobalt		Stains show:	None
carbonate	.4	Opacity:	Opaque
	100.0	Color	
		/oxidation:	Dark brown
		/reduction:	Dark brown

G 1078 Black

Colemanite	48		
Nepheline		Temperature:	C/09–06
syenite	29	Surface	
Zinc oxide	10	@ C/06:	Gloss
Black iron		Fluidity:	Fluid
oxide	5	Stains show:	No
Manganese		Opacity:	Opaque
dioxide	5	Color	
Cobalt oxide	3	/oxidation:	Black
	100	/reduction:	Black

G 1079 Whitish

Colemanite	40	Temperature:	C/08–06
Soda feldspar	35	Surface	
Barium		@ C/08:	Gloss
carbonate	10	Fluidity:	Little
Flint	10	Stains show:	Yes
Zinc oxide	5	Opacity:	Transparent
	100	Color	
		/oxidation:	Slight white
		/reduction:	Slight white
		Note:	Cracks

G 1080 Clear

Colemanite	45	Temperature:	C/08–07
Soda feldspar	40	Surface	
Barium		@ C/08:	Semi-gloss
carbonate	15	Fluidity:	Little
	100	Stains show:	Yes
		Opacity:	Transparent
		Color	
		/oxidation:	Clear
		Note:	Cracks

G 1081 Semi-Gloss White

Soda feldspar	28.80	Temperature:	C/08–06
Colemanite	26.25	Surface	
Frit #25	19.00	@ C/08:	Semi-gloss
Flint	9.00	Fluidity:	Little
Zinc oxide	8.35	Stains show:	Yes
Talc	3.50	Opacity:	Transparent
Whiting	3.00	Color	
Kaolin	2.10	/oxidation:	Slight white
	100.00	/reduction:	Slight white
		Note:	Cracks

G 1082 Tizzy

White lead	43.9	Temperature:	C/08–07
Frit #3134	16.7	Surface	
Flint	13.6	@ C/08:	Semi-gloss
Kaolin	12.3	Fluidity:	Little
Soda feldspar	8.8	Stains show:	Darks
Tin oxide	4.7	Opacity:	Translucent
	100.0	Color	
		/oxidation:	Broken whites
		/reduction:	Do not reduce
		Note:	Cracks

G 1083 White

Colemanite	68	Temperature:	C/08–07
Silica	19	Surface	
Kaolin	13	@ C/08:	Semi-gloss
	100	Fluidity:	Little
		Stains show:	Darks
		Opacity:	Opaque
		Color	
		/oxidation:	White
		/reduction:	White
		Note:	Slight bubbles where thick

G 1084 Leadless White
(Long Beach State, Long Beach, Ca.)

Sodium silicate, dry	54.5	Temperature:	C/08
Cryolite	19.5	Surface	
Flint	16.3	@ C/08:	Semi-matt
Bone ash	7.3	Fluidity:	Little
Bentonite	2.4	Stains show:	Darks
	100.0	Opacity:	Opaque
		Color	
		/oxidation:	White
		/reduction:	White
Copper oxide	3.4	/oxidation:	Turquoise

G 1085 Transparent Yellow

White lead	60.0	Temperature:	C/08
Flint	9.0	Surface	
Kaolin	7.9	@ C/08:	Gloss
Whiting	7.1	Fluidity:	Some
Zinc oxide	6.0	Stains show:	Yes
Soda feldspar	4.0	Opacity:	Transparent
Borax	4.0	Color	
Yellow stain	2.0	/oxidation:	Yellow
	100.0	/reduction:	Do not reduce
		Note:	Cracks

G 1086 Yellow

Colemanite	50	Temperature:	C/08
Boric acid	23	Surface	
Plastic vetrox	15	@ C/08:	Gloss
Uranium, yellow	10	Fluidity:	Some
Tin oxide	2	Stains show:	Yes
	100	Opacity:	Transparent
		Color	
		/oxidation:	Yellow
		/reduction:	Do not reduce

G 1087 Yellow Mustard

White lead	44.30	Temperature:	C/08
Flint	20.01	Surface	
Whiting	13.62	@ C/08:	Gloss
Kaolin	10.40	Fluidity:	Some
Tin oxide	5.46	Stains show:	Yes
Red iron		Opacity:	Transparent
oxide	3.65	Color	
Zinc oxide	2.56	/oxidation:	Yellow
	100.00		mustard
		/reduction:	Do not
			reduce

G 1088 Slight Yellow

Volcanic ash	30	Temperature:	C/08
Colemanite	25	Surface	
Whiting	25	@ C/08:	Gloss
Barium		Fluidity:	Little
carbonate	15	Stains show:	Yes
Zinc oxide	5	Opacity:	Transparent
	100	Color	
		/oxidation:	Slight
			yellow
		/reduction:	Clear
		Note:	Cracks

G 1089 Semi-Gloss Yellow

White lead	40	Temperature:	C/08–05
Soda feldspar	28	Surface	
Whiting	13	@ C/08:	Semi-gloss
Calcined		Fluidity:	Little
kaolin	8	Stains show:	Yes
Flint	5	Opacity:	Transparent
Zinc oxide	4	Color	
Zircopax	2	/oxidation:	Slight
	100		yellow
		/reduction:	Do not
			reduce

G 1090 Grayed Yellow

Red lead	30	Temperature:	C/08–06
Soda feldspar	30	Surface	
Colemanite	15	@ C/08:	Semi-gloss
Flint	12	Fluidity:	Little
Uranium		Stains show:	Darks
yellow	10	Opacity:	Translucent
Zinc oxide	3	Color	
	100	/oxidation:	Grayed
			yellow
		/reduction:	Grayed
			yellow
		Note:	Cracks

G 1091 Light Orange

Red lead	50	Temperature:	C/08
Colemanite	20	Surface	
Uranium		@ C/08:	Gloss
oxide	9	Fluidity:	Some
Soda feldspar	8	Stains show:	Yes
Yellow stain	5	Opacity:	Transparent
Tin oxide	4	Color	
Flint	4	/oxidation:	Light
	100		orange
		/reduction:	Do not
			reduce

G 1092 Orange

Colemanite	70.1	Temperature:	C/08
Plastic vetrox	16.2	Surface	
Uranium		@ C/08:	Gloss
oxide,		Fluidity:	Very fluid
yellow	6.3	Stains show:	Darks
Potassium		Opacity:	Translucent
nitrate	3.2	Color	
Vanadium	2.2	/oxidation:	Orange;
Zinc oxide	2.1		green
	100.1		where
			thin
		/reduction:	Red-
			orange;
			green
			where
			thin

G 1093 Coral Orange

Red lead	64	Temperature:	C/08
Lead		Surface	
chromate	22	@ C/08:	Gloss
Flint	11	Fluidity:	Some
Tin oxide	3	Stains show:	Darks
	100	Opacity:	Opaque
		Color	
		/oxidation:	Coral
			orange
		/reduction:	Do not
			reduce

G 1094 Red Raspberry

Red lead	30.3	Temperature:	C/08–06
Flint	21.4	Surface	
Whiting	15.1	@ C/06:	Gloss
Borax	16.0	Fluidity:	Fluid
Tin oxide	10.1	Stains show:	Darks
China clay	7.2	Opacity:	Translucent
Potassium		Color	
bichromate	1.0	/oxidation:	Red
	100.1		raspberry
		/reduction:	Do not
			reduce

G 1095 Red-Violet

Red lead	32.7	Temperature:	C/08–07
Flint	23.1	Surface	
Whiting	18.3	@ C/08:	Matt
Borax	17.4	Fluidity:	Little
Kaolin	4.9	Stains show:	Yes
Red stain	2.4	Opacity:	Translucent
Tin oxide	.6	Color	
Manganese		/oxidation:	Broken
carbonate	.4		red-violet
Cobalt		/reduction:	Broken
carbonate	.2		blue-violet
	100.0	Color	
		Note:	depends upon stain used

G 1096 Cobalt Blue

Colemanite	70	Temperature:	C/08–07
Silica	16	Surface	
Kaolin	13	@ C/08:	Gloss
Cobalt		Fluidity:	Little
carbonate	1	Stains show:	Yes
	100	Opacity:	Transparent
		Color	
		/oxidation:	Cobalt blue
		/reduction:	Cobalt blue
		Note:	Fine cracks

"Alligator on a Hat," Richard Shaw;
porcelain, stains, oxides, glazes;
8-3/4" H × 14-3/4" W.

G 1097 Blue

Kona feldspar	34	Temperature:	C/08
Colemanite	32	Surface	
Barium		@ C/08:	Gloss
carbonate	10	Fluidity:	Fluid
Flint	10	Stains show:	Yes
Lithium		Opacity:	Translucent
carbonate	6	Color	
Tin oxide	4	/oxidation:	Medium clean blue
Copper			
carbonate	4	/reduction:	Medium clean blue
	100		

G 1098 Green-Black

White lead	50.0	Temperature:	C/08
Soda feldspar	28.0	Surface	Gloss to
Whiting	9.0	@ C/08:	semi-gloss
Edgar plastic			
kaolin	5.0	Fluidity:	Some
Calcined		Stains show:	No
kaolin	4.0	Opacity:	Opaque
Flint	2.0	Color	
Copper		/oxidation:	Green-black
carbonate	1.0		
Chrome oxide	.7	/reduction:	Green-black
Cobalt			
carbonate	.3		
	100.0		

G 1099 Yellow-Green

Cullet	44.8	Temperature:	C/08–07
Colemanite	40.0	Surface	
China clay	7.6	@ C/08:	Gloss
Lead		Fluidity:	Fluid
chromate	7.6	Stains show:	Darks
	100.0	Opacity:	Translucent
		Color	
		/oxidation:	Medium yellow-green
		/reduction:	Lime green
		Note:	Cracks

G 1100 Olive Green

Kona feldspar	34	Temperature:	C/08–07
Colemanite	27	Surface	
Plastic vetrox	11	@ C/08:	Semi-gloss
Barium		Fluidity:	Little
carbonate	9	Stains show:	No
Flint	7	Opacity:	Opaque
Rutile	4	Color	
Yellow ochre	4	/oxidation:	Broken olive green with specks
Ilmenite	3		
Chrome oxide	1		
	100		

G 1101 Light Amber

Albany slip	45	Temperature:	C/08
Barium		Surface	
carbonate	15	@ C/08:	Gloss
Wollastonite	10	Fluidity:	Some
Frit #25		Stains show:	Yes
(Pemco)	10	Opacity:	Transparent
Lithium		Color	
carbonate	10	/oxidation:	Light
Flint	8		amber
Red iron		Note:	Cracks
oxide	2		
	100		

G 1102 Yellow Amber

White lead	49	Temperature:	C/08
Soda feldspar	18	Surface	
Flint	10	@ C/08:	Gloss
Whiting	10	Fluidity:	Some
Edgar plastic		Stains show:	Darks
kaolin	5	Opacity:	Translucent
Selenium	4	Color	
Calcined		/oxidation:	Yellow
kaolin	3		amber
Chrome oxide	1	/reduction:	Do not
	100		reduce
		Note:	Cracks

G 1103 Albany Slip Amber

Albany slip	30	Temperature:	C/08
Cornwall		Surface	Gloss to
stone	20	@ C/08:	semi-
Whiting	20		gloss
Lithium		Fluidity:	Some
carbonate	20	Stains show:	Darks
Boric acid	5	Opacity:	Translucent
Iron oxide	5	Color	
	100	/oxidation:	Yellow
			amber
		/reduction:	Rust brown

G 1104 Dark Gray

Cullet	50	Temperature:	C/08–06
Colemanite	15	Surface	
Ball clay	15	@ C/08:	Semi-gloss
Zinc oxide	12	Fluidity:	Little
Rutile	7	Stains show:	No
Cobalt		Opacity:	Opaque
carbonate	1	Color	
	100	/oxidation:	Dark gray
		/reduction:	Dark gray

G 1105 Soft Aik
(Long Beach State, Long Beach, Ca.)

Ground glass		Temperature:	C/08–06
(cullet)	25	Surface	
White lead	25	@ C/08:	Semi-gloss
Frit #25		Fluidity:	Little
(Pemco)	25	Stains show:	None
China clay	13	Opacity:	Opaque
Tin oxide	12	Color	
Setit	1	/oxidation:	Broken
	101		white
		/reduction:	Pepper
		Note:	Hold heat
			at C/08
			to break
			bubbles

G 1106 Dark Brown

White lead	50.03	Temperature:	C/08
Flint	16.00	Surface	
Soda feldspar	13.00	@ C/08:	Gloss
Colemanite	8.30	Fluidity:	Some
Manganese		Stains show:	No
dioxide	5.67	Opacity:	Opaque
Kaolin	4.33	Color	
Red iron		/oxidation:	Dark brown
oxide	2.67	/reduction:	Do not
	100.00		reduce

G 1107 Mustard Brown

White lead	46.2	Temperature:	C/08–07
Flint	19.2	Surface	
Tin oxide	11.5	@ C/08:	Gloss
Whiting	6.7	Fluidity:	Little
Edgar plastic		Stains show:	Darks
kaolin	5.8	Opacity:	Opaque
Calcined		Color	
kaolin	4.8	/oxidation:	Mustard
Red iron			brown
oxide	3.9	/reduction:	Do not
Zinc oxide	1.9		reduce
	100.0		

G 1108 Chocolate Ice Cream

Kona feldspar	30	Temperature:	C/08–06
Colemanite	25	Surface	
Plastic		@ C/08:	Semi-gloss
vetrox	15	Fluidity:	Little
Zinc oxide	9	Stains show:	No
Barium		Opacity:	Opaque
carbonate	7	Color	
Flint	5	/oxidation:	Rich
Burnt umber	5		chocolate
Whiting	3		ice cream
Chrome oxide	1	/reduction:	Rich
	100		chocolate
			ice cream

G 1109 Black

Calcined		Temperature:	C/08
borax	30	Surface	
Soda ash	25	@ C/08:	Gloss
Flint	25	Fluidity:	Some
Potassium		Stains show:	No
carbonate	8	Opacity:	Opaque
Red iron		Color	
oxide	6	/oxidation:	Black
Manganese		/reduction:	Black
dioxide	4		
Cobalt oxide	2		
	100		

G 1110 Leadless Clear

Colemanite	40	Temperature:	C/07–04
Boric acid	20	Surface	
Zinc oxide	20	@ C/04:	Gloss
Soda feldspar	15	Fluidity:	Fluid
Barium		Stains show:	Yes
carbonate	5	Opacity:	Transparent
	100	Color	
		/oxidation:	Clear
		/reduction:	Clear
		Note:	Cracks

G 1111 Slight White

Colemanite	56	Temperature:	C/07
Nepheline		Surface	
syenite	30	@ C/07:	Semi-gloss
Zinc	9	Fluidity:	Little
Zircopax	5	Stains show:	Yes
	100	Opacity:	Translucent
		Color	
		/oxidation:	Slight white
		/reduction:	Slight white
		Note:	Cracks

G 1112 Lukens Transparent Colemanite

Silica	33.2	Temperature:	C/07
Colemanite	30.7	Surface	
Soda feldspar	25.4	@ C/07:	Semi-gloss
Barium		Fluidity:	None
carbonate	10.7	Stains show:	Darks
	100.0	Opacity:	Translucent
		Color	
		/oxidation:	White
		/reduction:	White
		Note:	Cracks

Zircopax	14.0	=	Opaque white
Cobalt	2.0	=	Dark blue
Copper			
carbonate	3.2	=	Turquoise
Manganese			
carbonate	1.5	=	Gray-violet
Lead			
chromate	5.0	=	Yellow-green

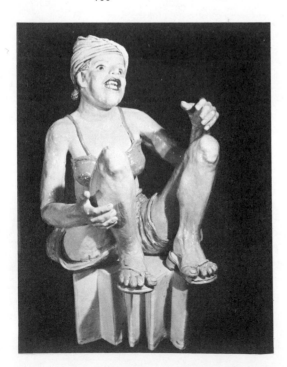

"Seated Lady," Viola Frey;
buff stoneware, C/01 white glaze,
C/05 overglaze; 33-3/4" H.

G 1113 Ochre Yellow

Kona feldspar	38
Colemanite	32
Uranium	10
Barium carbonate	9
Flint	9
Zinc oxide	1
Rutile	1
	100

Temperature:	C/07
Surface @ C/07:	Gloss
Fluidity:	Little
Stains show:	Yes
Opacity:	Transparent
Color /oxidation:	Light ochre yellow
/reduction:	Grayed yellow
Note:	Some cracks; grind the mixture

G 1114 Chinese Vermilion
(Long Beach State, Long Beach, Ca.)

White lead	64.8
Uranium, yellow	16.6
Flint	12.1
China clay	6.5
	100.0

Temperature:	C/07
Surface @ C/07:	Semi-gloss
Fluidity:	Fluid
Stains show:	Medium darks
Opacity:	Translucent
Color /oxidation:	Vermilion
Note:	Grind mixture

G 1115 Plum

Colemanite	55.9
Nepheline syenite	37.2
Tin oxide	4.9
Manganese carbonate	1.0
Cobalt carbonate	1.0
	100.0

Temperature:	C/07–05
Surface @ C/07:	Semi-gloss
Fluidity:	Little
Stains show:	No
Opacity:	Opaque
Color /oxidation:	Plum
/reduction:	Grayed plum

G 1116 Uranium Green

Kona feldspar	38
Colemanite	32
Barium carbonate	9
Flint	9
Uranium oxide	8
Tin oxide	2
Copper carbonate	2
	100

Temperature:	C/07–06
Surface @ C/07:	Gloss
Fluidity:	Little
Stains show:	Darks show
Opacity:	Translucent
Color /oxidation:	Medium grass green
/reduction:	Golds, oxblood red, green, copper
Note:	Some cracks; grind mixture

G1117 Blue-Green

Red lead	29.5
Flint	20.8
Whiting	16.5
Borax	15.6
China clay	8.3
Tin oxide	7.5
Antimony	1.1
Copper carbonate	.6
Cobalt carbonate	.1
	100.0

Temperature:	C/07
Surface @ C/07:	Gloss
Fluidity:	Little
Stains show:	None
Opacity:	Opaque
Color /oxidation:	Medium blue-green
/reduction:	Blue-green and luster
Note:	Apply thin coat; hold temperature to break bubbles

G 1118 Rutile Blue

White lead	40
Potash feldspar	30
Flint	10
Kaolin	10
Rutile	4
Tin oxide	3
Copper oxide	3
	100

Temperature:	C/07–04
Surface @ C/06:	Semi-gloss
Fluidity:	Little
Stains show:	No
Opacity:	Opaque
Color /oxidation:	Marbled jade
/reduction:	Broken tan

G 1119 Speckled Green

Soda feldspar	29
Red lead	28
Flint	18
Kaolin	14
Soda ash	7
Potassium bichromate	4
	100

Temperature:	C/07–04
Surface	
@ C/07:	Semi-gloss
@ C/04:	Gloss
Fluidity:	None
Stains show:	No
Opacity:	Opaque
Color	
/oxidation:	Speckled green
/reduction:	Deep gray-green

G 1120 Black

Nepheline syenite	30
Albany slip	20
Colemanite	20
Boric acid	19
Red iron oxide	10
Cobalt oxide	1
	100

Temperature:	C/07–06
Surface	
@ C/07:	Semi-gloss
Fluidity:	Little
Stains show:	No
Opacity:	Opaque
Color	
/oxidation:	Black
/reduction:	Black

(L) "Vampire Roller Derby Queen," (M) "I Am Sorry I Cannot Hear You— I Have a Banana in My Ear," (R) "Man on Roller Stilts," Bill Stewart; white earthenware, commercial over- and underglazes; 17" to 19" H.

G 1121 Clear

Frit #25		Temperature:	C/06–04
(Pemco)	45	Surface	
Barium		@ C/06:	Gloss
carbonate	20	Fluidity:	Little
Colemanite	20	Stains show:	Yes
Kaolin	6	Opacity:	Transparent
Flint	6	Color	
Zinc oxide	3	/oxidation:	Clear
	100	/reduction:	Clear; some pearling

G 1122 Leadless Clear

Flint	30	Temperature:	C/06
Soda feldspar	20	Surface	
Cryolite	20	@ C/06:	Gloss
Whiting	20	Fluidity:	Little
Lithium		Stains show:	Yes
carbonate	10	Opacity:	Transparent
	100	Color	
		/oxidation:	Clear
		/reduction:	Clear
		Note:	Cracks

G 1123 Lumpy Bumpy White

Flint	30.5	Temperature:	C/06–04
White lead	25.7	Surface	
Custer		@ C/06:	Gloss
feldspar	18.9	Fluidity:	Little
Borax	10.5	Stains show:	No
Pearl ash	5.4	Opacity:	Opaque
Kaolin	3.7	Color	
Whiting	3.6	/oxidation:	Lumpy bumpy white
Zinc oxide	1.7		
	100.0	/reduction:	Lumpy bumpy white
		Note:	Cracks

G 1124 Soft Gloss

Colemanite	60	Temperature:	C/06–04
Lepidolite	40	Surface	
	100	@ C/06:	Soft gloss
		Fluidity:	Little
		Stains show:	Yes
		Opacity:	Transparent
		Color	
		/oxidation:	Clear
		/reduction:	Clear
		Note:	Cracks

G 1125 Whitish

Colemanite	62	Temperature:	C/06
Flint	24	Surface	
Kaolin	14	@ C/06:	Semi-gloss
	100	Fluidity:	Some
		Stains show:	Yes
		Opacity:	Translucent
		Color	
		/oxidation:	Whitish
		/reduction:	Broken cool white
		Note:	Cracks

G 1126 Semi-Gloss

Flint	36.3	Temperature:	C/06
Borax	35.1	Surface	
Whiting	11.7	@ C/06:	Semi-gloss
Kaolin	6.5	Fluidity:	Some
Calcined		Stains show:	Darks
kaolin	6.5	Opacity:	Transparent
Soda feldspar	2.6		—thin;
Soda ash	1.3		opaque—
	100.0		thick
		Color	
		/oxidation:	White and clear
		/reduction:	White and clear

G 1127 Wood Ash

Colemanite	42.8	Temperature:	C/06–04
Soda feldspar	28.6	Surface	
Mixed wood		@ C/04:	Semi-gloss
ash	14.0	Fluidity:	Fluid
Plastic		Stains show:	Yes
vetrox	13.2	Opacity:	Translucent
Ball clay	1.4	Color	
	100.0	/oxidation:	Whitish
		/reduction:	Broken white and tan
		Note:	Cracks; color depends upon ash used

G 1128 White

Cullet	63	Temperature:	C/06–04
Zinc oxide	30	Surface	
Kaolin	7	@ C/06:	Satin matt
	100	Fluidity:	Fluid
		Stains show:	Darks
		Opacity:	Translucent
		Color	
		/oxidation:	White
		/reduction:	White
		Note:	Cracks

G 1129 Jenkins Eggshell Matt

Cullet	49.0	Temperature:	C/06–04
White lead	24.5	Surface	
China clay	14.7	@ C/06:	Semi-matt
Tin oxide	9.8	Fluidity:	Little
Setit powder	2.0	Stains show:	Darks
	100.0	Opacity:	Opaque
		Color	
		/oxidation:	Eggshell white
		/reduction:	Eggshell tan

G 1130 Clear Matt

Lithium carbonate	25	Temperature:	C/06–04
Colemanite	25	Surface	
Plastic vetrox	25	@ C/06:	Matt
Nepheline syenite	25	Fluidity:	Fluid
	100	Stains show:	Yes
		Opacity:	Transparent
		Color	
		/oxidation:	Clear
		/reduction:	Clear
		Note:	Cracks

G 1131 Bright True Yellow

Flint	36	Temperature:	C/06
Borax	36	Surface	
Uranium oxide, yellow	14	@ C/06:	Gloss
Colemanite	10	Fluidity:	Some
Boric acid	4	Stains show:	Yes
	100	Opacity:	Transparent
		Color	
		/oxidation:	Bright true yellow
		/reduction:	Do not reduce
		Note:	Some cracks; best on white clay

G 1132 Matt Yellow

White lead	59	Temperature:	C/06–04
Soda feldspar	19	Surface	
Barium carbonate	9	@ C/06:	Broken dry matt
Lead chromate	5	Fluidity:	None
Cornwall stone	4	Stains show:	Darks
China clay	4	Opacity:	Opaque
	100	Color	
		/oxidation:	Yellow— thin; orange— thick
		/reduction:	Grayed orange

G 1133 Orange Fleck

Borax	30	Temperature:	C/06
Soda feldspar	20	Surface	
Soda ash	16	@ C/06:	Gloss
Flint	15	Fluidity:	Some
Cadmium	6	Stains show:	Yes
Whiting	5	Opacity:	Transparent
Kaolin	4	Color	
Sulfur	3	/oxidation:	Clear with orange flecks
Selenium	1	/reduction:	Clear
	100	Note:	Cracks

G 1134 Yellow-Orange

Red lead	68	Temperature:	C/06
Flint	22	Surface	
Boric acid	4	@ C/06:	Gloss
Kaolin	3	Fluidity:	Little
Antimony	3	Stains show:	Yes
	100	Opacity:	Transparent
		Color	
		/oxidation:	Medium yellow-orange
		/reduction:	Ochre
		Note:	Cracks; use on white clay

G 1135 Broken Orange

White lead	65	Temperature:	C/06
Soda ash	12	Surface	
Flint	12	@ C/06:	Semi-gloss
Kaolin	8	Fluidity:	Little
Chrome oxide	3	Stains show:	Very darks
	100	Opacity:	Opaque
		Color	
		/oxidation:	Broken orange
		/reduction:	Broken browns
		Note:	Use on white clay

G 1136 Light Purple

Boric acid	47.0	Temperature:	C/06
Zircopax	19.5	Surface	
Soda ash	11.8	@ C/06:	Broken gloss and semi-matt
Flint	11.8		
Magnesium carbonate	5.9		
Cobalt carbonate	2.0	Fluidity:	Some
Manganese dioxide	2.0	Stains show:	No
	100.0	Opacity:	Opaque
		Color	
		/oxidation:	Broken light purple and tan
		/reduction:	Broken light purple and tan

G 1137 Turquoise Blue

Silicate of soda, dry	68.5	Temperature:	C/06–04
Borax	17.9	Surface	
Whiting	8.9	@ C/04:	Gloss
Copper carbonate	2.8	Fluidity:	Some
Bentonite	1.9	Stains show:	Yes
	100.0	Opacity:	Transparent
		Color	
		/oxidation:	Medium turquoise blue
		/reduction:	Blue, flashes of red, some luster

G 1138 Boron/Silicate Blue

Borax	35.2	Temperature:	C/06
Soda feldspar	29.4	Surface	
Flint	14.0	@ C/06:	Gloss
China clay	8.4	Fluidity:	Some
Whiting	7.5	Stains show:	Yes
Lithium carbonate	3.0	Opacity:	Transparent
Copper carbonate	2.0	Color	
Cobalt carbonate	.5	/oxidation:	Clean blue
	100.0	/reduction:	Clean blue
		Note:	Cracks

G 1139 Turquoise Green

Red lead	29.5	Temperature:	C/06–04
Flint	20.8	Surface	
Whiting	16.5	@ C/04:	Gloss
China clay	16.1	Fluidity:	Fluid
Borax	15.6	Stains show:	Yes
Copper carbonate	1.5	Opacity:	Transparent
	100.0	Color	
		/oxidation:	Turquoise green
		/reduction:	Do not reduce
		Note:	Cracks

G 1140 Persian Blue
(Long Beach State, Long Beach, Ca.)

Frit #25 (Pemco)	42.9	Temperature:	C/06
Flint	32.4	Surface	
China clay	13.0	@ C/06:	Semi-gloss
Tin oxide	9.1	Fluidity:	Little
Blue stain	2.6	Stains show:	No
	100.0	Opacity:	Opaque
		Color	
		/oxidation:	Persian blue
		/reduction:	Grayed Persian blue
		Note:	Color depends upon stain used

G 1141 Roc's Egg Glaze

Frit #2113	73.6	Temperature:	C/06
Soda feldspar	15.3	Surface	
Edgar plastic		@ C/06:	Semi-matt
kaolin	3.1	Fluidity:	Little
Zircopax	3.1	Stains show:	No
Rutile	1.5	Opacity:	Opaque
Copper		Color	
carbonate	.8	/oxidation:	Broken blues
Tin oxide	.4		
Manganese		/reduction:	Broken grayed blues
dioxide	.4		
Chrome			
oxide green	.4	Note:	Frit #2106 used in this test
	98.6		

G 1142 Blue-Black

Cullet	57.7	Temperature:	C/06
Zinc oxide	28.8	Surface	
Kaolin	6.8	@ C/06:	Matt
Nickel oxide	6.7	Fluidity:	Fluid
	100.0	Stains show:	No
		Opacity:	Opaque
		Color	
		/oxidation:	Blue-black
		/reduction:	Green, blue and black

G 1143 Jade

Borax	40	Temperature:	C/06–05
Zinc oxide	15	Surface	
Flint	15	@ C/06:	Gloss
Ball clay	15	Fluidity:	Little
Whiting	12	Stains show:	None
Copper		Opacity:	Opaque
carbonate	2	Color	
Chrome oxide	1	/oxidation:	Jade
	100	/reduction:	Dark jade, some luster
		Note:	Cracks

G 1144 Chrome Green

Soda feldspar	45.0	Temperature:	C/06–04
Soda ash	27.0	Surface	
Whiting	14.3	@ C/04:	Semi-gloss
Flint	10.0	Fluidity:	Little
Tin oxide	2.0	Stains show:	No
Chrome oxide	1.4	Opacity:	Opaque
Cobalt		Color	
carbonate	.3	/oxidation:	Warm chrome green
	100.0		
		/reduction:	Cool chrome green

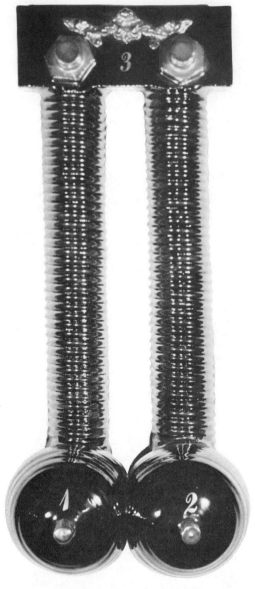

"Mechanical Breast Form," Victor Spinski;
stoneware, glaze, gold luster; 31-1/2" H.

G 1145 Lithium Green

Flint	40.8	Temperature:	C/06
Cryolite	14.3	Surface	
Soda feldspar	14.2	@ C/06:	Semi-gloss
Lithium		Fluidity:	Some
carbonate	9.6	Stains show:	No
Bone ash	8.3	Opacity:	Opaque
Whiting	6.4	Color	
China clay	3.8	/oxidation:	Dark green
Copper		/reduction:	Dark green
carbonate	2.0		
Chrome oxide	.4		
Cobalt			
carbonate	.2		
	100.0		

G 1146 Grass Green

Cullet	60.9	Temperature:	C/06–04
Zinc oxide	29.0	Surface	
Kaolin	6.8	@ C/06:	Semi-matt
Nickel oxide	3.3	Fluidity:	None
	100.0	Stains show:	No
		Opacity:	Opaque
		Color	
		/oxidation:	Dark olive
		/reduction:	Do not reduce

G 1147 Dark Olive

Red lead	52.8	Temperature:	C/06–05
Flint	20.2	Surface	
China clay	15.2	@ C/06:	Matt
Rutile	10.2	Fluidity:	Little
Copper		Stains show:	No
carbonate	1.6	Opacity:	Opaque
	100.0	Color	
		/oxidation:	Dark olive
		/reduction:	Do not reduce

G 1148 Rutile and Ilmenite

Borax	35	Temperature:	C/06
Spodumene	30	Surface	
Kaolin	7	@ C/06:	Gloss
Whiting	7	Fluidity:	Some
Flint	6	Stains show:	Darks
Rutile	6	Opacity:	Opaque
Ilmenite	4	Color	
Lithium	3	/oxidation:	Broken tan, ochre with specks
Burnt umber	2		
	100	/reduction:	Broken tan, ochre with specks

G 1149 Tan

Barium		Temperature:	C/06–05
carbonate	50.0	Surface	
Boric acid	17.0	@ C/06:	Semi-gloss
Nepheline		Fluidity:	Little
syenite	10.0	Stains show:	Yes
Kaolin	7.0	Opacity:	Transparent
Plastic		Color	
kaolin	7.0	/oxidation:	Slight tan
Flint	4.0	/reduction:	Slight tan
Opax	2.5		
Cerium oxide	1.5		
Vanadium	1.0		
	100.0		

G 1150 Warm Gray

Colemanite	50	Temperature:	C/06
Albany slip	21	Surface	
Plastic		@ C/06:	Gloss
vetrox	21	Fluidity:	Fluid
Red iron		Stains show:	No
oxide	8	Opacity:	Opaque
	100	Color	
		/oxidation:	Warm gray
		/reduction:	Dark gray

G 1151 Slight Gray

Barium		Temperature:	C/06
carbonate	51.6	Surface	
Boric acid	19.4	@ C/06:	Matt
Nepheline		Fluidity:	Fluid
syenite	11.8	Stains show:	Darks
Kaolin	8.6	Opacity:	Translucent
Flint	5.4	Color	
Copper		/oxidation:	Slight gray
carbonate	1.9		
Nickel oxide	1.3		
	100.0		

G 1152 Leadless Brown

Calcined		Temperature:	C/06–04
borax	30	Surface	
Potassium		@ C/05:	Matt
carbonate	20	Fluidity:	None
Soda ash	15	Stains show:	No
Red iron		Opacity:	Opaque
oxide	15	Color	
Kaolin	10	/oxidation:	Rich medium broken browns
Zinc oxide	10		
	100		
		/reduction:	Dark broken browns

G 1153 Clear

Lead		Temperature:	C/05
bisilicate	55	Surface	
Boric acid	25	@ C/05:	Gloss
Talc	10	Fluidity:	Some
Zinc oxide	10	Stains show:	Yes
	100	Opacity:	Transparent
		Color	
		/oxidation:	Clear
		/reduction:	Clear
		Note:	Cracks

G 1154 Leadless Clear

Kona feldspar	25	Temperature:	C/05–04
Whiting	25	Surface	
Barium		@ C/04:	Gloss
carbonate	25	Fluidity:	Fluid
Colemanite	25	Stains show:	Yes
	100	Opacity:	Transparent
		Color	
		/oxidation:	Clear
		/reduction:	Clear
		Note:	Cracks

G 1155 Slight White

Colemanite	67	Temperature:	C/05–04
Plastic		Surface	
vetrox	25	@ C/05:	Semi-gloss
Zircopax	8	Fluidity:	None
	100	Stains show:	Yes
		Opacity:	Translucent
		Color	
		/oxidation:	Slight white
		/reduction:	Slight white
		Note:	Fine cracks

G 1156 Whitish

Kona feldspar	23	Temperature:	C/05–03
Whiting	23	Surface	
Barium		@ C/05:	Semi-gloss
carbonate	23	Fluidity:	Little
Colemanite	23	Stains show:	Yes
Zircopax	8	Opacity:	Translucent
	100	Color	
		/oxidation:	Whitish
		/reduction:	Whitish
		Note:	Cracks

G 1157 Cullet White

Cullet	70	Temperature:	C/05–04
Soda feldspar	18	Surface	
Kaolin	6	@ C/04:	Semi-gloss
Flint	6	Fluidity:	Little
	100	Stains show:	Darks
		Opacity:	Translucent
		Color	
		/oxidation:	Whitish
		/reduction:	Whitish
		Note:	Apply even coat

G 1158 White

Cullet	70	Temperature:	C/05–04
Soda feldspar	15	Surface	
Ball clay	10	@ C/05:	Semi-matt
Flint	5	Fluidity:	None
	100	Stains show:	Very darks
		Opacity:	Opaque
		Color	
		/oxidation:	White
		/reduction:	White
		Note:	Apply even coat

G 1159 Mulberry

White lead	41.3	Temperature:	C/05–04
Flint	26.6	Surface	
Soda feldspar	14.9	@ C/05:	Semi-gloss
Zinc oxide	4.3	Fluidity:	Little
Manganese		Stains show:	No
carbonate	4.0	Opacity:	Opaque
Tin oxide	3.2	Color	
China clay	3.0	/oxidation:	Mulberry
Whiting	2.7	/reduction:	Dark mulberry
	100.0		

G 1160 Lithium Blue

Borax	31	Temperature:	C/05
Soda feldspar	19	Surface	
Flint	19	@ C/05:	Gloss
Soda ash	11	Fluidity:	Some
Lithium		Stains show:	Yes
carbonate	9	Opacity:	Transparent
Kaolin	5	Color	
Whiting	4	/oxidation:	Lithium blue
Copper			
carbonate	2	/reduction:	Lithium blue
	100	Note:	Cracks

G 1161 Turquoise Blue

Cullet	64.4	Temperature:	C/05–04
Soda feldspar	14.4	Surface	
Borax	5.2	@ C/05:	Semi-gloss
Zinc oxide	3.6	Fluidity:	Little
Barium		Stains show:	Darks
carbonate	3.6	Opacity:	Translucent
Flint	2.7	Color	
Copper		/oxidation:	Turquoise
carbonate	2.7		blue
Ball clay	1.8	/reduction:	Turquoise
Bentonite	1.6		green and
	100.0		copper
			luster

G 1162 Leadless Blue

Lepidolite	36.9	Temperature:	C/05
Flint	29.1	Surface	
Lithium		@ C/05:	Semi-matt
carbonate	19.4	Fluidity:	Some
Talc	4.9	Stains show:	Yes
Barium		Opacity:	Transparent
carbonate	4.9	Color	
Copper		/oxidation:	Thin =
carbonate	2.9		light blue
Bentonite	1.9		turquoise
CMC glue	.2		Thick =
	100.2		dark blue
			turquoise
		/reduction:	Broken
			blue and
			copper
			luster

G 1163 Taffy Tan

Albany slip	35	Temperature:	C/05–04
Colemanite	35	Surface	
Nepheline		@ C/05:	Satin gloss
syenite	20	Fluidity:	Fluid
Plastic		Stains show:	Darks
vetrox	7	Opacity:	Translucent
Red iron		Color	
oxide	3	/oxidation:	Taffy tan
	100	/reduction:	Dark taffy
			tan
		Note:	Cracks;
			apply
			medium
			thickness

G 1164 Rust Brown

Lead		Temperature:	C/05–04
bisilicate	75.9	Surface	
Local red		@ C/05:	Gloss
clay	16.6	Fluidity:	Some
Tin oxide	2.8	Stains show:	Yes
Red iron		Opacity:	Translucent
oxide	2.8	Color	
Yellow ochre	1.9	/oxidation:	Rich rust
	100.0		brown
		/reduction:	Tan brown

G 1165 Rutile Brown

Red lead	40	Temperature:	C/05–04
Flint	25	Surface	
Rutile	15	@ C/04:	Matt
Potash		Fluidity:	Some
feldspar	10	Stains show:	Darks
Whiting	10	Opacity:	Opaque
	100	Color	
		/oxidation:	Warm
			brown
		/reduction:	Do not
			reduce

"Church" (lidded candle holder),
William Wilhelmi;
earthenware, natural, glazed; 12" H.

G 1166 Clear

Flint	30	Temperature:	C/04–03
Wollastonite	26	Surface	
Cryolite	15	@ C/04:	Gloss
Lithium		Fluidity:	None
carbonate	15	Stains show:	Yes
Kaolin	7	Opacity:	Transparent
Zinc oxide	7	Color	
	100	/oxidation:	Clear
		/reduction:	Clear
		Note:	Cracks

G 1167 Transparent

Lead		Temperature:	C/04
bisilicate	61.2	Surface	
Nepheline		@ C/04:	Semi-gloss
syenite	28.5	Fluidity:	Little
Whiting	9.4	Stains show:	Yes
Bentonite	.9	Opacity:	Transparent
	100.0	Color	
		/oxidation:	Clear, some pearling
		/reduction:	Clear, some pearling
		Note:	Cracks

G 1168 White

da feldspar	19.7	Temperature:	C/04
nt	16.1	Surface	
hiting	14.5	@ C/04:	Semi-gloss
Borax	12.4	Fluidity:	Some
Kaolin	11.3	Stains show:	Yes
Zircopax	10.8	Opacity:	Translucent
Barium		Color	
carbonate	6.8	/oxidation:	White
Zinc oxide	6.8	/reduction:	White
Talc	1.6		
	100.0		

G 1169 Opaque White

Flint	25	Temperature:	C/04
Borax,		Surface	
calcined	19	@ C/04:	Semi-gloss
Potassium		Fluidity:	Little
carbonate	19	Stains show:	Darks
Soda ash	16	Opacity:	Opaque
Alumina	9	Color	
Barium		/oxidation:	White
carbonate	5	/reduction:	White
Whiting	4	Note:	Cracks;
Tin oxide	3		grind
	100		mixture

G 1170 Frost Matt

Soda feldspar	35	Temperature:	C/04–03
Flint	20	Surface	
Cryolite	15	@ C/04:	Frost matt
Whiting	15	Fluidity:	None
Soda ash	15	Stains show:	Darks
	100	Opacity:	Opaque
		Color	
		/oxidation:	White
		/reduction:	White
		Note:	Cracks

"Sculptural Form," Ralph Bacerra; stoneware, matt glaze, luster, overglaze; 48-1/2" H.

G 1171 Slight Yellow

Red lead	63.2	Temperature:	C/04–03
Flint	22.5	Surface	
China clay	12.5	@ C/04:	Gloss
Soda feldspar	1.6	Fluidity:	Little
Soda ash	.2	Stains show:	Yes
	100.0	Opacity:	Transparent
		Color	
		/oxidation:	Slight yellow
		/reduction:	Slight yellow
		Note:	Cracks

G 1172 Golden Yellow

Red lead	61.7	Temperature:	C/04
Flint	22.2	Surface	
China clay	12.2	@ C/04:	Gloss
Red iron		Fluidity:	Fluid
oxide	2.6	Stains show:	Yes
Soda feldspar	1.3	Opacity:	Transparent
	100.0	Color	
		/oxidation:	Golden yellow
		/reduction:	Cool yellow
		Note:	Cracks; bubbles where thick

G 1173 Ram Bam
(Long Beach State, Long Beach, Ca.)

White lead	47.7	Temperature:	C/04
Flint	14.0	Surface	
Kona F-4	11.7	@ C/04:	Broken gloss and matt
Kaolin	8.8		
Barium			
carbonate	8.1	Fluidity:	Some
Whiting	4.2	Stains show:	Yes
Maine feldspar		Opacity:	Translucent
(plastic		Color	
vetrox used		/oxidation:	Light lemon yellow
in test)	3.5		
Rutile	2.0	/reduction:	Do not reduce
	100.0	Note:	Cracks

G 1174 Naples Yellow

Lead		Temperature:	C/04–03
bisilicate	60.1	Surface	
Flint	17.7	@ C/04:	Semi-gloss
Colemanite	6.1	Fluidity:	Little
Tin oxide	5.6	Stains show:	Darks
Kaolin	4.6	Opacity:	Opaque
Lithium		Color	
carbonate	3.7	/oxidation:	Naples yellow
Antimony			
oxide	2.3	/reduction:	Do not reduce
	100.1		
		Note:	Bubbles in thick areas

G 1175 Bright Yellow-Orange

White lead	42.5	Temperature:	C/04
Soda feldspar	16.2	Surface	
Whiting	11.1	@ C/04:	Matt
Pink stain		Fluidity:	No
(Leslie		Stains show:	Darks
#421)	10.1	Opacity:	Translucent
Flint	10.1	Color	
Kaolin	5.0	/oxidation:	Bright yellow-orange
Antimony			
oxide	3.0		
Tin oxide	2.0	/reduction:	Subtle yellow-orange
	100.0		
		Note:	Color depends upon stain used

G 1176 Chrome Red and Orange

White lead	65.0	Temperature:	C/04
Flint	16.1	Surface	
Soda feldspar	10.0	@ C/04:	Gloss
Kaolin	4.6	Fluidity:	Fluid
Chrome oxide	2.5	Stains show:	None
Soda ash	1.8	Opacity:	Opaque
	100.0	Color	
		/oxidation:	Broken dark red and orange
		/reduction:	Do not reduce

"Beaverjack," Viola Frey;
buff stoneware, C/01 white glaze. C/015 overglazes;
41-1/4" H.

G 1177 Grayed Purple

Colemanite	58.8	Temperature:	C/04–03
Kaolin	22.0	Surface	
Silica	18.0	@ C/04:	Semi-gloss
Manganese		Fluidity:	Little
carbonate	1.0	Stains show:	Yes
Cobalt		Opacity:	Transparent
carbonate	.2		—thin;
	100.0		translu-
			cent—
			thick
		Color	
		/oxidation:	Grayed
			purple
		/reduction:	Grayed
			purple
		Note:	Cracks

G 1178 Blue Ink

Red lead	61.6	Temperature:	C/04
Flint	21.0	Surface	
China clay	11.0	@ C/04:	Gloss
Soda feldspar	3.2	Fluidity:	Fluid
Copper		Stains show:	Darks
carbonate	2.4	Opacity:	Translucent
Cobalt oxide	.8	Color	
	100.0	/oxidation:	Medium
			blue ink
		/reduction:	Metallic
			black
		Note:	Cracks;
			bubbles
			where
			thick

G 1179 Gray-Blue

Soda feldspar	42.9	Temperature:	C/04
Soda ash	19.0	Surface	
Colemanite	19.0	@ C/04:	Semi-matt
Whiting	16.0	Fluidity:	Little
Tin oxide	1.9	Stains show:	Darks
Red iron		Opacity:	Translucent
oxide	.9	Color	
Cobalt		/oxidation:	Pale light
carbonate	.3		gray-blue
	100.0	/reduction:	Pale soft
			gray-blue
		Note:	Cracks

G 1180 Jade Green

Soda feldspar	39	Temperature:	C/04
Colemanite	35	Surface	
Barium		@ C/04:	Semi-gloss
carbonate	10	Fluidity:	Some
Cornwall		Stains show:	No
stone	8	Opacity:	Opaque
Potassium		Color	
dichromate	5	/oxidation:	Jade green
Zircopax	3	/reduction:	Jade green
	100		

G 1181 Yellow-Green

Frit #25		Temperature:	C/04
(Pemco)	38	Surface	
Flint	29	@ C/04:	Gloss
Uranium		Fluidity:	Fluid
oxide,		Stains show:	Yes
yellow	14	Opacity:	Transparent
Lithium		Color	
carbonate	12	/oxidation:	Light
Magnesium			yellow-
carbonate	3		green
Zinc oxide	3	/reduction:	Do not
Bentonite	1		reduce
	100	Note:	Cracks

"Ceramic Form," Ka-Kwong Hui;
ceramic whiteware, overglaze, luster;
39-1/2" H × 20" W.

G 1182 Cream Ivory

White lead	38	Temperature:	C/04
Flint	15	Surface	
Zinc oxide	15	@ C/04:	Semi-matt
Cullet	15	Fluidity:	None
Edgar plastic		Stains show:	Darks
kaolin	5	Opacity:	Opaque
Ball clay	5	Color	
Tin oxide	4	/oxidation:	Cream
Titanium	3		ivory
	100	/reduction:	Cream
			ivory

G 1183 Barnard Clay Slip Glaze
(Sylvia Hyman, Nashville, Tenn.)

Albany clay	40	Temperature:	C/04–1
White lead	40	Surface	
Barnard clay	20	@ C/04:	Gloss
	100	Fluidity:	None
		Stains show:	Darks
		Opacity:	Opaque
		Color	
		/oxidation:	Rich amber
		/reduction:	Rich amber
			brown

G 1184 True Black

White lead	47.40	Temperature:	C/04–03
Ball clay	15.89	Surface	
Zircopax	10.28	@ C/04:	Semi-gloss
Flint	8.28	Fluidity:	Little
Whiting	6.00	Stains show:	No
Bentonite	5.61	Opacity:	Opaque
Red iron		Color	
oxide	3.27	/oxidation:	True black
Copper oxide	2.80		
Cobalt oxide	.47		
	100.00		

G 1185 Matt Black

Local red		Temperature:	C/04–02
clay	40.4	Surface	
Litharge	35.9	@ C/04:	Dry matt
Red iron	8.7	Fluidity:	None
Manganese		Stains show:	No
carbonate	6.7	Opacity:	Opaque
Silica	6.0	Color	
Whiting	2.3	/oxidation:	Black
	100.0	/reduction:	Do not
			reduce

Figure 2.1

"Segments," John W. Conrad; melted glass in stoneware; 22" square.

two

glass and clay

Today almost every aspect of human endeavor comes into direct or indirect contact with glass. The earliest known use of glass, according to archaeologists, occurred about 6,000 years ago: during the Early Dynastic period of Egypt, a glasslike covering was used for ceramic forms. The use of glass by itself has been traced to Mesopotamia around 1600 B.C. Slender rods of colored glass were softened so that they could be twisted around a clay core to make a bottle form. The object was heated, the rods fused together, the core removed, and the glass polished to produce a multicolored bottle. During this period the combining of glass and clay was developed. Clay balls were made into bead forms and heated; then the surface was decorated with strings, dots, shapes, and ribbons of softened glass. The surface was smoothed by flame polishing. These attractive, portable ornaments, an excellent medium of trade with other countries, have been found in Italy and other countries in Europe, in China, and in Indonesia.

The use of glass with clay in contemporary creative ceramic arts involves primarily the melting of glass in recessed areas of clay. Frits, jars, bottles, plate glass, window glass, and other glass objects are crushed, then placed in the bottom of ceramic dishes, ashtrays, tile, and other forms. Glazes cannot achieve the low cost, bright colors, three-dimensional depth, and crackle effect of fractured glass. (A good example is fried marbles.) Many interesting effects, colors, and textures can be achieved only by using glass with clay. Figure 2.1 is an example of melted glass in clay tile framed in wood.

Glass as a Glaze

One important use of glass is as a glaze. The composition of glazes and glasses is very similar. Glass can be made inexpensively into *cullet* (ground glass) for a glaze. Almost any glass—clear, translucent, opaque, colorless, or colored—can be crushed, placed into a ball-mill and ground, washed to remove impurities, and screened through a 200-mesh screen. Cullet and ground glass come directly from glass; a *frit* is a compound made from a glaze, which can be used like ground glass for the melted glass process. Colorants can be added to ground glass, frit, or cullet, using the percentages specified in glaze formulas. The results will be similar to glazes. A small amount of bentonite or glue added to ground glass acts as a binder to hold the glass to the surface of bisqueware. As with all frit- (cullet-, ground glass–) based glazes, adjustments of other glaze materials (clay, fluxes, opacifiers, matting agents) may be needed to make up a completed glaze compound. Red, orange, yellow, and other unusually colored glass can be a source of colors to make a glaze not achievable by any other method. For example, ruby red glass makes a red C/6 reduced glaze.

Glass Fused to Clay

I became interested in stained glass effects and began combining colored glass with clay, a technique in which glass became part of the structure instead of an additive decoration. The goal was to use the clay structure for strength, form, and support for the glass, while taking advantage of the color and transparency of glass. After seeing several of the resulting sculptures, which used glass cemented to an opening in clay, a friend asked if it would be possible to fuse glass to clay. His intriguing question was the stimulus for a long-term research project.

I tested every possible commercial glass and clay to obtain a glass-and-clay combination in which the coefficient of expansion/contraction would be the same in both materials. Even if the glass fused to the clay in the cooling cycle, most clay and glass shrank at

different rates, and either the clay or the glass would crack and the sculpture fall apart. The only solution was to formulate clay and glass that were compatible, a feat that required hundreds of tests. At long last, success: color plate 1 shows a piece in which glass, used like a stained glass window, is fused across an opening in clay.

The clay form is made and bisque-fired; the glaze (if any) is applied; and the piece is high-fired. Then the individually made glass, cut to shape, is set in place and the piece is fired for the third time to the point that the glass slumps to fit the contour and fuse to the clay (fig. 2.2). The kiln is turned off and the blowers kept on to reduce the temperature about 50 degrees, preventing residual heat from further softening the glass. The dampers then are closed and the kiln cools slowly. Thin-walled kilns that cool very quickly are not recommended.

Figure 2.2

Cross section of fusing glass to clay:
(A) opening in clay form;
(B) glass set in place;
(C) glass fused to clay.

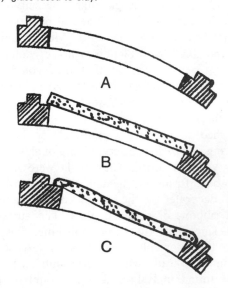

Melted Glass in Clay

The use of melted glass has not been fully explored, although several ceramists have recently worked with melted glass in clay as a unique means of enriching clay form. They use readily available glass, commercial cullet, enamel, and frit, and even formulate new glass to be used on tile, the sides of pots, sculpture, container lids, and the edges of bowls.

CRUSHING GLASS

A word of caution: smashing or crushing glass is dangerous because of flying particles. Be extremely careful to prevent flying glass and pieces of glass lying around the studio. Most glass products are too big or difficult to use in the recessed areas of your clay form and therefore must be diminished in size. Use a piece of heavy cloth as an envelope for the glass and strike it with a hammer. Alternatively, place the glass between several layers of newspaper and break it with a hammer, remove the glass, and throw away the newspaper with glass fragments embedded in it. Or, place the glass in a warm kiln or oven. When the glass is 400° F. or warmer, remove it and quickly immerse it in a bucket of water to shatter. No matter which method you use, *be careful*. Store crushed glass in heavy paper bags or similar heavy-duty containers, and label them.

GLASS COLORANTS

You can obtain unusual colors and effects in glass; however, you must do copious testing and research to develop a variety. Obtaining certain colors in glass is an exacting process, just as it is with glazes, for a colorant will often produce a particular color in one glass composition but a very different color in another.

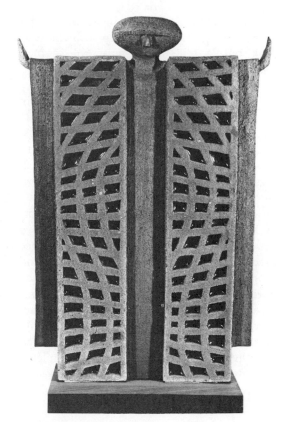

"Rejoice," John W. Conrad;
melted glass in stoneware, natural, stain;
24"H × 16"W × 2-1/2"D.

Numerous variables affect glass and colorants. Various glass bases, such as potash, soda-potash-lead, lead, soda, or boron, will influence the absorption spectra of the manganese in glass. An alkali replaced by another of a higher atomic weight will sometimes cause the color to shift from red-purple to blue-purple. The variables that influence glass during melting and cooling are the temperature at which the glass is melted, the atmospheric conditions, the cooling rate, the composition of the glass, and the character of the kiln's atmosphere. The composition of the clay also affects colorants. These variables create conditions in which manganese in the glass may result in such colors as pink, purple, blue-purple, red-purple, garnet, green, blue, brown, yellow (lithium-lead-silicate base), or gray.

To obtain a particular color with a certain colorant requires a particular glass formula and related conditions. For the ceramist this means a knowledge of many different glass formulas, an extensive chemical inventory, and familiarity with chemistry. In our lab/studio we ran several extensive test series to achieve unusual colors and effects by mixing colorants and opacifiers with a raw glass compound. The testing sequence used the formula MG 17, compounded and weighed into 50-gram portions. Into each portion the various colorants were dry-mixed and the mixture was placed in bisque-fired porcelain cups, then in a gas-fired kiln that was fired to 2150° F. in a slightly reduced atmosphere. Specimens of the completed melted glass tests, showing some possible colors, are illustrated in figure 2.3.

Figure 2.3

Specimens of completed melted glass tests.

COLORANTS FOR MELTED GLASS

Colorant	Percent	Resulting Color
Arsenic Chrome oxide	2.0 .1	Light sea green
Arsenic oxide Copper carbonate	4.0 3.0	Dark olive
Burnt sienna	3.0	Green-black
Burnt umber	2.0	Slight beer-bottle brown
Cadmium sulfide Zinc oxide Chrome oxide	3.0 .6 .1	Light true green
Cadmium sulfide Zinc oxide Cobalt carbonate	2.8 .6 .1	Light blue-gray
Cadmium sulfide Zinc oxide	4.0 .6	Gray-brown
Cerium oxide Manganese carbonate	2.0 .4	Light grayed yellow
Cerium oxide Titanium oxide	5.0 2.0	Beer-bottle brown
Cerium oxide Vanadium	1.0 .8	Gray-green
Chrome oxide	.4	Light green
Chrome oxide Iron oxide, red	.3 .2	Light green
Cobalt carbonate	.03	Slight blue
Cobalt carbonate	.07	Medium blue
Copper carbonate	.6	Light turquoise
Copper carbonate	1.2	Dark turquoise
Copper carbonate Chrome oxide	.4 .2	Light green-turquoise
Iron oxide, red	4.0	Black
Iron oxide, red	1.0	Light moss green
Iron oxide, red Cobalt carbonate	.4 .02	Gray
Iron oxide, red Copper carbonate	.8 .4	Olive
Iron oxide, red Nickel carbonate	.2 .2	Amber
Kryolith Copper phosphate	6.0 2.0	Kelly green with flashes of red
Lead chromate	1.6	Bright green
Lithium carbonate Nickel carbonate	5.0 .2	Amber

Colorant	Percent	Resulting Color
Magnesium carbonate Iron oxide, red Cobalt carbonate	.8 .4 .1	Steel blue
Manganese carbonate	2.0	Amethyst
Manganese carbonate	5.0	Dark amethyst
Manganese carbonate Cobalt carbonate	1.60 .08	Blue
Manganese carbonate Copper carbonate	2.0 .6	Warm green
Nickel carbonate	1.0	Cool tan
Potassium dichromate	.4	Bright green
Rutile	2.0.	Tan
Tin oxide	10.0	White
Tin oxide	5.0	Milky white
Tin oxide Copper carbonate	.5 .4	Gray
Titanium oxide Copper carbonate	2.0 .4	Deep blue
Titanium oxide Antimony	2.0 1.0	Slight yellow
Titanium dioxide Uranium oxide, yellow	2.0 1.4	Slight yellow
Vanadium pentoxide	.6	Dark gray-olive
Vanadium pentoxide Chrome oxide	.8 .1	Green
Yellow ochre	.8	Tan
Yellow stain Cobalt carbonate	4.00 .04	Blue-green
Zinc oxide Cobalt carbonate	1.00 .03	Slight blue

Obvious sources of glass for the melted glass technique are window and container glass. Other interesting colors result from using commercial cullet, enamel, frit, raw chemicals, and raw glass formulas. To illustrate the possible varieties, we obtained or made up several different glasses and subjected them to a uniform testing sequence. We made porcelain cups that would hold 50 grams of powdered glass, cullet, crushed glass, or mixed raw glass. We placed the glass in these bisque-fired porcelain cups, put them in a gas-fired kiln, and fired it to 2150° F. in a slightly reduced atmosphere. Then we turned off the kiln and allowed it to cool at its own rate. We noted the resulting color, opacity, surface, cracks, and unusual effects. Most of the glass cracked, from 0 to over 100 cracks per square inch. To standardize the amount of crackage in the cooled glass, we used a scale of 1 to 5 (see figure 2.4).

The chart of glass testings (fig. 2.5) is organized into the following categories: commercial cullet; enamel; frit; bottle, window, and container glass; raw chemical formulas; and raw glass formulas. For each formula the chart gives data about its surface, crackle scale, opacity, resulting color, and any unusual effects.

Figure 2.4

Melted glass crackle scale.

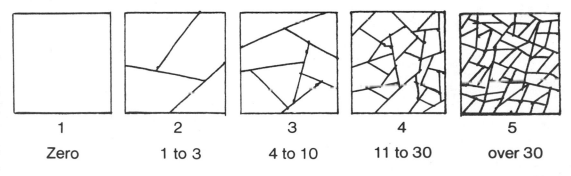

1	2	3	4	5
Zero	1 to 3	4 to 10	11 to 30	over 30

Figure 2.5 (*pages 48–54*)

Glass testings. Explanation : *Surface* refers to the surface
quality of gloss, semi-gloss, semi-matt, or matt
at the temperature fired. *Crackle scale* refers to the
standard 1–5 numbered scale of crackage in the glass.
Transparency refers to the degree of transparency
of the glass : opaque, translucent, or transparent.
Temperature refers to the temperature tested.
Fired color is the color resulting from a slight reduction firing.

Commercial Cullet
(Available from Keystone Cullet Co., 426 Willow Crossing Rd., Greensburg, Pa.)

Keystone #186 Lead		Keystone #190 Lead	
Surface:	Gloss	Surface:	Gloss
Crackle scale:	5	Crackle scale:	5
Transparency:	Transparent	Transparency:	Transparent
Temperature:	C/5	Temperature:	C/5
Fired color:	Blue-green	Fired color:	Grayed light green

Keystone #95 Lime		Keystone #105 Lime	
Surface:	Gloss	Surface:	Gloss
Crackle scale:	5	Crackle scale:	5
Transparency:	Transparent	Transparency:	Transparent
Temperature:	C/5	Temperature:	C/5
Fired color:	Medium blue	Fired color:	Garnet

Keystone #3		Keystone #8 Lime	
Surface:	Gloss	Surface:	Gloss
Crackle scale:	4	Crackle scale:	5
Transparency:	Opaque and transparent areas	Transparency:	Transparent
Temperature:	C/5	Temperature:	C/5
Fired color:	Orange	Fired color:	Yellow-green

Keystone #9 Lime		Keystone #189 Lead	
Surface:	Gloss	Surface:	Semi-gloss
Crackle scale:	5	Crackle scale:	5
Transparency:	Transparent	Transparency:	Transparent
Temperature:	C/5	Temperature:	C/5
Fired color:	Red	Fired color:	Light purple

Keystone #183 Lead		Keystone #115 Lime	
Surface:	Gloss	Surface:	Gloss
Crackle scale:	5	Crackle scale:	5
Transparency:	Transparent	Transparency:	Transparent
Temperature:	C/5	Temperature:	C/5
Fired color:	Yellow diamond	Fired color:	Clear

Keystone #94 Lime

Surface:	Gloss
Crackle scale:	5
Transparency:	Transparent
Temperature:	C/5
Fired color:	Slight blue-green

Keystone #14 Lime

Surface:	Gloss
Crackle scale:	4
Transparency:	Opaque
Temperature:	C/5
Fired color:	White

Keystone #106

Surface:	Gloss
Crackle scale:	5
Transparency:	Transparent
Temperature:	C/5
Fired color:	Tan

Keystone #107 Lime

Surface:	Gloss
Crackle scale:	5
Transparency:	Transparent
Temperature:	C/5
Fired color:	Light yellow

Keystone #97 Lime

Surface:	Gloss
Crackle scale:	5
Transparency:	Transparent
Temperature:	C/5
Fired color:	Grayed light olive

Keystone #61 Lime

Surface:	Gloss
Crackle scale:	5
Transparency:	Opaque
Temperature:	C/5
Fired color:	Broken white, gray, and orange

Keystone #96 Lime

Surface:	Gloss
Crackle scale:	5
Transparency:	Transparent
Temperature:	C/5
Fired color:	Gray

Keystone #104 Lime

Surface:	Gloss
Crackle scale:	5
Transparency:	Transparent
Temperature:	C/5
Fired color:	Light turquoise

Keystone #2 Lime

Surface:	Gloss
Crackle scale:	5
Transparency:	Opaque
Temperature:	C/5
Fired color:	Milky

Keystone #92 Lime

Surface:	Gloss
Crackle scale:	5
Transparency:	Transparent
Temperature:	C/5
Fired color:	Champagne

Keystone #59 Lime

Surface:	Gloss
Crackle scale:	5
Transparency:	Transparent
Temperature:	C/5
Fired color:	Light green

Keystone #184 Lead

Surface:	Gloss
Crackle scale:	3
Transparency:	Transparent
Temperature:	C/2
Fired color:	Gray

Keystone #260 (26% Lead)

Surface:	Gloss
Crackle scale:	5
Transparency:	Transparent
Temperature:	C/5
Fired color:	Light gray

Keystone #110 Lime

Surface:	Semi-gloss
Crackle scale:	3
Transparency:	Opaque
Temperature:	C/2
Fired color:	White

Keystone #56 Lime		Keystone #57 Lime	
Surface:	Gloss	Surface:	Gloss
Crackle scale:	5	Crackle scale:	3
Transparency:	Transparent	Transparency:	Transparent
Temperature:	C/5	Temperature:	C/2
Fired color:	Clear	Fired color:	Dark brown

Keystone #58 Lime		Keystone #613C Lime	
Surface:	Gloss	Surface:	Gloss
Crackle scale:	5	Crackle scale:	3
Transparency:	Transparent	Transparency:	Transparent
Temperature:	C/5	Temperature:	C/2
Fired color:	Medium green	Fired color:	Beer-bottle brown

Keystone #84 Lime		Keystone #77 Lime	
Surface:	Gloss	Surface:	Gloss
Crackle scale:	4	Crackle scale:	3
Transparency:	Opaque	Transparency:	Opaque
Temperature:	C/2	Temperature:	C/2
Fired color:	Broken medium greens	Fired color:	Broken blues

Keystone #180 Lead		Keystone #198 Lime	
Surface:	Gloss	Surface:	Gloss
Crackle scale:	5	Crackle scale:	3
Transparency:	Transparent	Transparency:	Opaque
Temperature:	C/2	Temperature:	C/2
Fired color:	Clear	Fired color:	Light green

Keystone #1 Lime		Keystone #4–4A Lime (Cadmium selenium)	
Surface:	Gloss	Surface:	Gloss
Crackle scale:	3	Crackle scale:	3
Transparency:	Opaque	Transparency:	Transparent
Temperature:	C/2	Temperature:	C/2
Fired color:	Broken yellow	Fired color:	Tan

Keystone #89 Lime		Keystone #181 Lead	
Surface:	Gloss	Surface:	Gloss
Crackle scale:	5	Crackle scale:	3
Transparency:	Opaque	Transparency:	Opaque
Temperature:	C/2	Temperature:	C/2
Fired color:	White	Fired color:	Black

Keystone #52 Lime		Keystone #86 Lime	
Surface:	Gloss	Surface:	Gloss
Crackle scale:	4	Crackle scale:	3
Transparency:	Opaque	Transparency:	Opaque
Temperature:	C/2	Temperature:	C/2
Fired color:	Slight green	Fired color:	White, pink, and black

Keystone #87A Lime

Surface:	Gloss
Crackle scale:	4
Transparency:	Opaque
Temperature:	C/2
Fired color:	Broken greens

Keystone #103

Surface:	Gloss
Crackle scale:	3
Transparency:	Opaque
Temperature:	C/2
Fired color:	Red

Keystone #170 Lime

Surface:	Semi-gloss
Crackle scale:	3
Transparency:	Opaque
Temperature:	C/2
Fired color:	Light green

Keystone #7 Lime

Surface:	Gloss
Crackle scale:	3
Transparency:	Transparent
Temperature:	C/2
Fired color:	Medium blue

Keystone #5–12 Lime

Surface:	Gloss
Crackle scale:	3
Transparency:	Transparent
Temperature:	C/2
Fired color:	Clear

(Available from Talmach Supply, P.O. Box 2332, La Puente, Ca. 91746)

Talmach Crystal

Surface:	Gloss
Crackle scale:	5
Transparency:	Transparent
Temperature:	C/5
Fired color:	Clear

Talmach Opal

Surface:	Gloss
Crackle scale:	4
Transparency:	Translucent
Temperature:	C/5
Fired color:	Milky

Talmach Regular

Surface:	Gloss
Crackle scale:	5
Transparency:	Transparent
Temperature:	C/5
Fired color:	Slight green

Talmach Flint

Surface:	Gloss
Crackle scale:	5
Transparency:	Transparent
Temperature:	C/5
Fired color:	Clear

Enamel
(Available from T. C. Thompson Enamels, 1539 Old Deerfield Rd., Highland Park, Ill.)

Thompson Enamel #961

Surface:	Gloss
Crackle scale:	5
Transparency:	Opaque
Temperature:	C/1
Fired color:	Dark blue-black

Thompson Enamel #986

Surface:	Gloss
Crackle scale:	5
Transparency:	Transparent
Temperature:	C/1
Fired color:	Medium yellow-green

Frit

Frit #311

Surface:	Semi-gloss
Crackle scale:	3
Transparency:	Transparent
Temperature:	C/6
Fired color:	Clear

Frit #3230

Surface:	Gloss
Crackle scale:	5
Transparency:	Transparent
Temperature:	C/5
Fired color:	Clear

Frit #626

Surface:	Gloss
Crackle scale:	3
Transparency:	Transparent
Temperature:	C/6
Fired color:	Clear

Frit #14

Surface:	Gloss
Crackle scale:	5
Transparency:	Transparent
Temperature:	C/5
Fired color:	Slight green

Frit #85

Surface:	Gloss
Crackle scale:	2
Transparency:	Transparent
Temperature:	C/6
Fired color:	Yellow

Frit #399

Surface:	Gloss
Crackle scale:	1
Transparency:	Opaque
Temperature:	C/5
Fired color:	Slight yellow-green

Frit #28

Surface:	Gloss
Crackle scale:	2
Transparency:	Transparent
Temperature:	C/7
Fired color:	Slight yellow

Frit #2106

Surface:	Semi-gloss
Crackle scale:	2
Transparency:	Opaque
Temperature:	C/06
Fired color:	White

Frit #25

Surface:	Semi-gloss
Crackle scale:	3
Transparency:	Translucent
Temperature:	C/06
Fired color:	Whitish

Frit #740

Surface:	Matt
Crackle scale:	1
Transparency:	Opaque
Temperature:	C/04
Fired color:	Slight yellow

Bottle, Window, and Container Glass

Coors Beer Bottle

Surface:	Gloss
Crackle scale:	5
Transparency:	Transparent
Temperature:	C/5
Fired color:	Beer-bottle brown

Clear Window Glass

Surface:	Gloss
Crackle scale:	5
Transparency:	Transparent
Temperature:	C/5
Fired color:	Clear

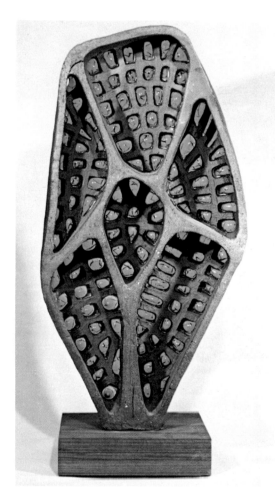

"Gainsborough Portrait Jar," Tyrone Larson;
stoneware, glaze, decal, luster;
12" H × 8-1/4" W.

"Segments Six," John W. Conrad;
fused glass to stoneware, natural;
28" H × 15" W × 3-3/4".

"Covered Box," John Glick;
stoneware, glazes, slip, oxides; 4" H × 11" × 7".

"Vase," Tom Turner;
white stoneware, vapor glazed with salt
and copper fumed; 9-1/2" H.

"The Little Known Birds of the Inner Eye,"
Rhoda Le Blanc Lopez;
stoneware, natural; 24" H × 24" diam.

"Pedestaled Compote," Julie Larson;
stoneware, white glaze, mother-of-pearl
and platinum luster;
9" H × 6-1/4" W.

"Bottle," Dorothy Bearnson; stoneware,
gloss glaze, luster, gold leaf; 9-1/2" H × 8" diam.

"Rhino Buggy," Bill Stewart;
white earthenware, commercial under- and overglazes: 21-1/2" H.

"Log Hatchet Bottle," Richard Shaw;
porcelain, glazes, stains, oxides; 17-1/4" H.

"Two Narrow-Neck Bottles," Rose Cabat;
porcelain, crystalline; 10-1/2" H. 3-3/4" H.

"Third Steven," Susan Kemenyffy and Steven Kemenyffy;
Raku, low-temperature glazes, lusters; 22-1/2" diam.

"Covered Jar," Steve McGovney; porcelain, matt blue-green; 18'' H.

"Feather Fetish #3-4," Ken Shores; feathers, stoneware, glaze, natural; 17-1/2'' H × 12-1/2'' square.

"Medusa Series II, III, IV," Helen Watson; stoneware, oxide rub, matt glaze; 12'' H, 20'' H, and 12-1/2'' H.

"Porcelain Bottle," Jack Feltman;
porcelain, peach-cream reduction crystal glaze;
14'' H × 5-1/2'' square.

"White Tea Bowl," Marcia Manhart;
talc clay, low-fire glaze and luster;
6-3/4" H × 6-1/4" W.

Mickeys Beer Bottle (Green)

Surface:	Gloss
Crackle scale:	5
Transparency:	Transparent
Temperature:	C/5
Fired color:	Medium green

Green Window Glass

Surface:	Gloss
Crackle scale:	5
Transparency:	Transparent
Temperature:	C/5
Fired color:	Grayed medium green

Yellow-Green Wine Bottle

Surface:	Gloss
Crackle scale:	5
Transparency:	Transparent
Temperature:	C/5
Fired color:	Grayed yellow-green

Noxzema Jar

Surface:	Gloss
Crackle scale:	5
Transparency:	Transparent where thin
Temperature:	C/5
Fired color:	Medium cobalt blue

Green Wine Bottle

Surface:	Gloss
Crackle scale:	5
Transparency:	Transparent
Temperature:	C/5
Fired color:	Medium green

(Available from Blinko Glass Co., Milton, W. Va. 25541)

Blinko Turquoise

Surface:	Gloss
Crackle scale:	5
Transparency:	Transparent
Temperature:	C/5
Fired color:	Turquoise

Blinko Cobalt Blue

Surface:	Gloss
Crackle scale:	5
Transparency:	Transparent
Temperature:	C/5
Fired color:	Cobalt blue

Blinko Ruby Red

Surface:	Gloss
Crackle scale:	5
Transparency:	Transparent
Temperature:	C/5
Fired color:	Dark brown

Blinko Orange

Surface:	Gloss
Crackle scale:	5
Transparency:	Transparent
Temperature:	C/5
Fired color:	Orange

Raw Chemical

Bicarbonate of Soda

Surface:	Gloss
Crackle scale:	5
Transparency:	Transparent
Temperature:	C/6
Fired color:	Slight yellow
Note:	Boils during firing

Colemanite

Surface:	Gloss
Crackle scale:	3
Transparency:	Transparent
Temperature:	C/02
Fired color:	Slight yellow

Borax

Surface:	Gloss
Crackle scale:	5
Transparency:	Transparent
Temperature:	C/04
Fired color:	Clear
Note:	Boils during firing

Ground Glass

Surface:	Matt
Crackle scale:	3
Transparency:	Opaque
Temperature:	C/6
Fired color:	White

Boric Acid

Surface:	Gloss
Crackle scale:	4
Transparency:	Transparent
Temperature:	C/04
Fired color:	Clear
Note:	Some boiling during firing

Sodium Silicate, Dry #SS65

Surface:	Semi-gloss
Crackle scale:	4
Transparency:	Translucent
Temperature:	C/6
Fired color:	Slight white

Glass Formulas

MG 164

Whiting	35	Surface:	Gloss
Colemanite	30	Crackle scale:	5
Kaolin	15	Transparency:	Transparent
Flint	10	Temperature:	C/5
Magnesium carbonate	10	Fired color:	Slight greenish
	100		

MG 5

Flint	25	Surface:	Gloss
Whiting	23	Crackle scale:	3
Magnesium carbonate	20	Transparency:	Transparent
Kaolin	16	Temperature:	C/8
Boric acid	12	Fired color:	Slight greenish
Soda ash	4		
	100		

MG 163

Whiting	25	Surface:	Gloss
Boric acid	20	Crackle scale:	1
Flint	20	Transparency:	Transparent
Magnesium carbonate	20	Temperature:	C/5
Kaolin	15	Fired color:	Clear
	100		

MG 1

Kaolin	28	Surface:	Gloss
Flint	27	Crackle scale:	3
Whiting	24	Transparency:	Transparent
Boric acid	12	Temperature:	C/6
Magnesium carbonate	5	Fired color:	Slight greenish
Soda ash	4		
	100		

The combination of clay and transparent glass can produce exciting color, texture, and decorative effect. Imaginative ceramists have used chunks of glass on the inside of planters and bowls; on the lips of dishes and plates; on the outside of bottles, jars, vases, and sculpture; on plates and tile; and on recessed areas of lids, plaques, and handles.

You can create an interesting effect by placing chunks of glass on the sides of a ceramic form that has been bisque-fired and glazed and is ready to be high-fired. The glass will melt and flow down the sides of the form, making a smooth and glossy surface in contrast with the glaze; yet the base glaze will be visible through the glass. Do not use too much glass on the side of the form, or the fired glass will flow down the side and fuse the bottom of the form to the kiln shelf. This sticking problem can occur if the glass is over-fired, even if the correct amount of glass was used. The hotter the glass, the more fluid it becomes, so be careful to control the temperature correctly. Clay protrusions or cups will hold the chunks of glass on the sides of the clay form during firing and will become a decorative part of the surface design (fig. 2.6).

Most glass softens about 1500° F. At 1900° and above the glass acquires a smooth, even low with a glossy, bubble-free finish. Both the glass melting and the glaze firing can be done at the same time. The amount of glass placed in the recessed areas is a main concern; the pressure of too much glass may crack the entire clay form. A glass thickness of approximately one-quarter inch is sufficient to display the color and transparency. To avoid overflow, place crushed, powdered, or raw glass compound in the recessed areas to an estimated level of three-quarters of the recessed area when the glass is melted (fig. 2.7). Capillary action will cause the melted glass to flow up the sides of the recessed area.

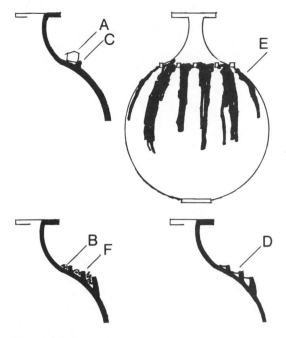

Figure 2.6

Use of glass on sides of clay forms:
(A) chunk of glass; (B) crushed glass;
(C) clay projection to hold glass in place;
(D) clay cup form to hold glass;
(E) fired form showing glass flow;
(F) glass melted into cup form.

Figure 2.7

Cross section of clay forms showing glass placement (left) and melted glass results (right):
(A) powdered glass; (B) crushed glass;
(C) fired glass showing capillary action.

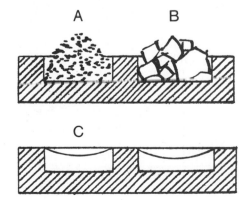

"Raku Vase," Richard Wukich;
stoneware clay, Rakued, matt glazes; 15-1/4" H.

three

raku

Raku, from the Chinese word for enjoyment, has many different connotations. It is the name of a ceramic technique, an object, a philosophical state of mind, and a religious endeavor. Unfortunately, the word *Raku* has become associated with instant, quick-fired vessels, a concept almost opposite from the traditional Japanese idea of appreciation of materials, design, tactile sensibility, and total simplicity. Raku is more than instant ceramics; it may include any of the following: low-fire procedure, ware cooled in water, ware reduced in sawdust, ware fired in a preheated kiln, ware made into a teabowl form, or ware made from highly grogged earthenware clay.

The earliest use of the Raku process—removing hot ware from the kiln to cool quickly—has not been determined, though it seems to have originated in the late Muromachi period in Japan (fourteenth to sixteenth centuries). During this time extensive importing of Chinese ceramics into Japan stimulated Japanese potteries to new growth and development. Chinese pottery, fashionable and respected, was used for tea drinking. Marato Shuko (1422–1503), a Buddhist monk, set formal rules of etiquette for the Japanese tea ceremony. The ceremony created a special relationship between the host and the guest, emphasizing the drinking of tea, the tea ceremony utensils, and the appreciation of the traits of humility and simplicity. The ceramics for the ceremony had quiet beauty and utility; the aesthetics of the clay form were to be unassuming and somber. About 1572 the first Raku tea ceremony utensils were produced by Chojiro, who had originally made roof tiles using the Raku process.

The shapes, colors, textures, and glazes have changed slightly over the years, but the concept of restrained teabowls and other tea equipment from the earth has remained part of the charm of Raku. Raku, with its rich historic background, can offer insight to the ceramist and widen his consciousness of creative ceramics.

The Raku process is the rapid firing and cooling of a clay form. The glaze and clay used are not unusual. A typical Raku object is made of a heavily grogged, open clay body, using hand formation or wheel-thrown techniques. Because of the nature of the clay and the heavy-handed treatment, the clay wall is usually thicker than it is in most equivalent ceramic forms. Potters have developed varying approaches to the Raku process. When it is placed in the kiln, the piece may be wet, bone dry, or bisqued for the Raku firing. For surface treatment the piece may be glazed, burnished, or left natural. The kiln may be hot or cold when the clay piece is inserted. When the glaze turns glossy—from 1400° to 1900° F., depending upon the glaze—the piece is removed from the kiln to cool quickly, either by quenching in water or by placement into a container of organic materials to burn and reduce. Sometimes it is set on a brick to cool.

Clay Bodies

Raku forms can be made of almost any clay body that can take the thermal shock of fast heating and cooling. Generally, the clay has a high content of grog, fireclay, or other coarse-grained materials to give it openness (porosity) and strength. The amount of grog (small, medium, or large-sized) usually ranges from 5 to 40 percent. The inclusion of 3 to 10 percent talc will help the clay resist thermal shock. Fireclay content ranges up to 50 percent.

The shape and size of the ceramic form are factors in the choice of the clay body. A small bowl, three inches by two inches, could be made with a porcelain throwing clay. However, a large, heavy jar, sixteen inches by twelve inches, requires a heavily grogged clay. Most commercial or individually made clay bodies will work satisfactorily with the addition of a coarse fireclay and grog.

The color of the clay body does not matter for reduction because the reducing process will turn the clay charcoal, gray, or black. In oxidation firing, various colors obtained with ceramic pigments give the clay body color.

Bisque-Firing

There are several approaches to the art of Raku firing. Bisque-firing the ceramic forms is necessary for most clay bodies and techniques. It is possible to make a ceramic form, apply a slip glaze, and set it in a hot kiln while it is still wet. A protective envelope of steam surrounds it as it dries. After it dries completely, it fires as a regular glaze firing. This wet firing process requires a very "open," heavily grogged clay so that the moisture can evaporate quickly. Otherwise the loss factor of such a process can be very high.

Because the typical Raku piece is fragile, consider bisque-firing at higher-than-usual temperatures. Bisque-firing is typically done about 1500° F., but if the temperature is raised to 1900° F. the clay will be stronger and will retain the ability to withstand the thermal shocks of the Raku process. Stoneware-type clays require higher bisque-firing temperatures than earthenware and Raku clays.

Raku Glazes

A major appeal of Raku, together with its simplicity of design, is the use of uncomplicated glazes. Most Raku glazes are composed of a limited number of glaze materials including lead, soda, boron, flint, clay, and frits. Many formulas contain only two to five ingredients in addition to the colorants, yet they create a variety of glaze effects and a wide range of colors. The firing temperature is generally between 1450° and 1950° F.

Because the classical Japanese Raku glazes have a high lead content, they are used today primarily for nonfood ware. Many ceramists have eliminated raw lead from their glazes because of concern for public health and desire to comply with government regulations. High schools and colleges sometimes eliminate raw lead (white lead, red lead, and litharge) from their inventory to prevent the accidental use of lead in glazes. Raw lead is not necessary for Raku glazes, for a sufficient number of individually and commercially made lead-based frits, nonlead frits, and glazes is available. (See the section on lead in glazes in chapter 1.)

Glaze formulas for the Raku process differ very little from those listed as low-temperature glazes in chapter 1. Process and attitude make Raku unique, not materials. The formulas listed below were subjected to the same testing procedures as the low-temperature glazes.

TESTING

We used porcelain clay to make "L"-shaped Raku testing forms such as those in figure 1.1. A black engobe was painted on one edge of the greenware, which was then bisque-fired to 1900° F. The glazes were weighed in 100-gram units, mixed with water to a painting consistency, and applied with a brush to two test pieces. Both were placed in an electric kiln and fired to the suggested temperature. One of the test pieces was then removed from the kiln, placed in a lidded metal container filled with newspaper, and allowed to cool to about 600°. It was then removed from the container and set on a brick to cool to finger-handling temperature. The other test piece was removed at the same time and placed on an asbestos pad to air-cool to finger-handling temperature. Some of the Raku testings are shown in figure 3.1.

After the glazes cooled, we analyzed their quality and characteristics using a uniform procedure. We recorded the following information for each formula: (1) ability of the glaze to fuse to the clay; (2) development of a smooth glaze surface; (3) degree of transparency; (4) surface texture; (5) glaze flow; (6) temperature firing range; and (7) reduction and oxidation colors of the glaze. The formulas listed below are organized into gloss, semi-gloss, semi-matt, and matt surfaces and further subdivided as to transparent, translucent, and opaque qualities; then they are distinguished by color: colorless, white, yellow, red-orange, pink, blue, green, tan, and brown.

Figure 3.1

Raku testings.

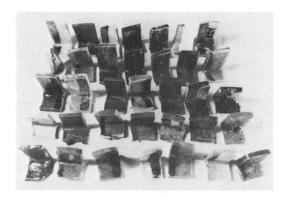

Raku Glaze Formulas

R 101 Leadless Clear

Colemanite	84	Temperature:	C/08–05
Kentucky ball		Surface	Gloss and
#4	8	@ C/08:	semi-gloss
Tin oxide	8	Fluidity:	Some
	100	Stains show:	Yes
		Opacity:	Transparent
		Color	
		/oxidation:	Clear
		/reduction:	Clear
		Note:	Some cracks

R 102 Clear Lead

White lead	50	Temperature:	C/08–06
Flint	29	Surface	
Calcined		@ C/08:	Gloss
borax	17	Fluidity:	Little
Kaolin	4	Stains show:	Yes
	100	Opacity:	Transparent
		Color	
		/oxidation:	Clear
		/reduction:	Clear
		Note:	Cracks

R 103 Lead/Colemanite

Colemanite	43.6	Temperature:	C/08–06
White lead	34.5	Surface	
Frit #170	9.1	@ C/08:	Gloss
Flint	9.1	Fluidity:	Little
Zircopax	3.7	Stains show:	Yes
	100.0	Opacity:	Transparent
		Color	
		/oxidation:	Clear
		/reduction:	Clear
		Note:	Cracks

R 104 Clear Crackle

White lead	40	Temperature:	C/010–08
Kaolin	20	Surface	
Colemanite	20	@ C/010:	Gloss
Frit #25	10	Fluidity:	Little
Flint	10	Stains show:	Yes
	100	Opacity:	Transparent
		Color	
		/oxidation:	Clear
		/reduction:	Clear
		Note:	Cracks

"Partly Cloudy" (landscape container), Wayne Higby; earthenware, Rakued, Raku glazes; 13" H × 13-1/4" W.

R 105 Leadless Clear Crackle

Colemanite	40	Temperature:	C/09–08
Whiting	16	Surface	
Plastic vetrox	12	@ C/09:	Gloss
Barium		Fluidity:	Some
carbonate	8	Stains show:	Yes
Nepheline		Opacity:	Transparent
syenite	8	Color	
Lithium		/oxidation:	Clear
carbonate	8	/reduction:	Clear
Vanadium		Note:	Cracks
pentoxide	8		
	100		

R 106 Leadless Frit

Frit #5301	79.2	Temperature:	C/010–08
Borax	9.9	Surface	
Edgar plastic		@ C/010:	Semi-gloss
kaolin	5.0	Fluidity:	Some
Tin oxide	5.0	Stains show:	Yes
Bentonite	.9	Opacity:	Translucent
	100.0	Color	
		/oxidation:	Slight whitish
		/reduction:	Slight whitish

R 107 Colemanite

Colemanite	60
Soda feldspar	20
Cullet	20
	100

Temperature: C/09–07
Surface
@ C/09: Semi-gloss
Fluidity: Little
Stains show: Yes
Opacity: Translucent
Color
/oxidation: Milky
/reduction: Milky
Note: Cracks

R 108 Cullet-Colemanite

Cullet	40
Colemanite	40
Flint	10
Nepheline syenite	10
	100

Temperature: C/08–06
Surface
@ C/08: Semi-gloss
Fluidity: Little
Stains show: Yes
Opacity: Translucent
Color
/oxidation: Whitish
/reduction: Whitish

R 109 Raku 333

Colemanite	33.3
Nepheline syenite	33.3
Plastic vetrox	33.3
	99.9

Temperature: C/04
Surface
@ C/04: Semi-gloss
Fluidity: Little
Stains show: Darks
Opacity: Translucent
Color
/oxidation: Milky
/reduction: Milky
Note: Occasional cracks

"Triangle Springs" (landscape container), Wayne Higby; earthenware, Rakued, Raku glazes; 10-1/2" H × 6-3/4" W.

Oxide	Percent	Oxidation Color	Reduction Color
Cobalt	1.4	Dark blue	Dark blue
Copper carbonate	.8	Light turquoise	Grayed turquoise
Iron oxide red	4.0	Grayed tan	Grayed tan
Manganese carbonate	3.2	Medium purple-brown	Medium purple-brown
Nickel oxide	6.0	Medium grayed green	Grayed olive
Tin oxide	8.0	White	White
Zircopax	8.0	White	White
Black iron oxide	5.0 ⎫		
Manganese dioxide	5.0 ⎬	Black	Black
Cobalt oxide	3.0 ⎭		

R 110 Leadless White

Colemanite	51.0	Temperature:	C/07–04
Kaolin	34.3	Surface	
Flint	14.7	@ C/07:	Semi-gloss
	100.0	Fluidity:	Little
		Stains show:	Darks
		Opacity:	Translucent
		Color	
		/oxidation:	White
		/reduction:	White

R 111 Crackle White

Colemanite	48	Temperature:	C/04
Soda feldspar	25	Surface	
Barium		@ C/04:	Semi-gloss
carbonate	12	Fluidity:	Little
Zircopax	10	Stains show:	Darks
Flint	5	Opacity:	Opaque
	100	Color	
		/oxidation:	White
		/reduction:	White
		Note:	Some cracks

R 112 Leadless White Crackle

Colemanite	40	Temperature:	C/05–04
Nepheline		Surface	
syenite	40	@ C/05:	Semi-gloss
Soda feldspar	14	Fluidity:	Little
Zircopax	6	Stains show:	Darks
	100	Opacity	Opaque
		Color	
		/oxidation:	White
		/reduction:	White
		Note:	Cracks

R 113 Uranium Yellow

Colemanite	59	Temperature:	C/08–06
Plastic vetrox	11	Surface	
Barium		@ C/08:	Gloss
carbonate	11	Fluidity:	Some
Nepheline		Stains show:	Yes
syenite	11	Opacity:	Transparent
Uranium,		Color	
yellow	8	/oxidation:	Yellow
	100	/reduction:	Yellow
		Note:	Cracks; grind mixture thoroughly

R 114 Chromate Yellow

White lead	40.8	Temperature:	C/010–08
Frit #3124	40.8	Surface	
Flint	10.2	@ C/010:	Gloss
Edgar plastic		Fluidity:	Some
kaolin	5.1	Stains show:	Yes
Lead		Opacity:	Transparent
chromate	3.1	Color	
	100.0	/oxidation:	Yellow
		/reduction:	Dark yellow
		Note:	Cracks

R 115 Lead Yellow

Potassium		Temperature:	C/08
nitrate	35	Surface	
Red lead	35	@ C/08:	Gloss
Flint	15	Fluidity:	Fluid
Boric acid	10	Stains show:	Yes
Flint	5	Opacity:	Transparent
	100	Color	
		/oxidation:	Yellow
		/reduction:	Grayed yellow
		Note:	Many cracks

R 116 Fluid Yellow

Red lead	33	Temperature:	C/08
Flint	33	Surface	
Potash		@ C/08:	Gloss
nitrate	33	Fluidity:	Fluid
Glue	1	Stains show:	Yes
	100	Opacity:	Transparent
		Color	
		/oxidation:	Yellow
		/reduction:	Grayed yellow
		Note:	Many cracks

R 117 Red-Orange

Red lead	60	Temperature:	C/07
Whiting	13	Surface	
Flint	10	@ C/07:	Matt
Soda feldspar	6	Fluidity:	Some
China clay	4	Stains show:	Darks
Chromium		Opacity:	Opaque
oxide	4	Color	
Tin oxide	3	/oxidation:	Red-orange
	100	/reduction:	Warm dark brown

R 118 Pink

Frit #W-18 (lead)	70	Temperature:	C/06–03
Tin oxide	15	Surface	
China clay	10	@ C/04:	Semi-gloss
Flint	5	Fluidity:	Some
	100	Stains show:	Darks
		Opacity:	Opaque
		Color	
		/oxidation:	Pale pink, white, or light lavender
		/reduction:	Grayed white

R 119 Leadless Turquoise

Frit #14	72.6	Temperature:	C/08
Flint	10.4	Surface	
Soda ash	10.4	@ C/08:	Gloss
Colemanite	5.2	Fluidity:	Some
Copper carbonate	1.4	Stains show:	Yes
	100.0	Opacity:	Transparent
		Color	
		/oxidation:	Turquoise
		/reduction:	Dark turquoise with some copper luster
		Note:	Cracks

R 120 Cobalt Blue

Colemanite	70	Temperature:	C/08–06
Borax	10	Surface	
Zircopax	10	@ C/08:	Gloss
Ball clay	9	Fluidity:	Some
Cobalt carbonate	1	Stains show:	Darks
	100	Opacity:	Translucent = thin; opaque = thick
		Color	
		/oxidation:	Light blue
		/reduction:	Medium blue

R 121 Gloss Turquoise

Colemanite	51	Temperature:	C/08–06
Plastic vetrox	21	Surface	
Cullet	13	@ C/08:	Gloss
Nepheline syenite	11	Fluidity:	Little
Copper carbonate	4	Stains show:	Yes
	100	Opacity:	Translucent
		Color	
		/oxidation:	Turquoise
		/reduction:	Copper luster

R 122 Lithium Blue

Colemanite	50	Temperature:	C/017–014
Soda feldspar	25	Surface	
Lithium carbonate	23	@ C/017:	Matt
Copper carbonate	1	Fluidity:	Little
Tin oxide	1	Stains show:	Yes
	100	Opacity:	Transparent
		Color	
		/oxidation:	Medium turquoise
		/reduction:	Turquoise
		Note:	Mix well and apply evenly

R 123 Copper Luster

Frit #25	40.0	Temperature:	C/08–06
Colemanite	35.5	Surface	
Borax	10.0	@ C/08:	Gloss
Flint	10.0	Fluidity:	Some
Red copper oxide	2.5	Stains show:	Yes
Tin oxide	2.0	Opacity:	Transparent
	100.0	Color	
		/oxidation:	Light turquoise
		/reduction:	Green with red copper luster
		Note:	Cracks

R 124 Lead Green

White lead	89.5	Temperature:	C/018–017
Copper carbonate	4.6	Surface	
Flint	3.2	@ C/018:	Gloss
Frit #2106	2.7	Fluidity:	Fluid
	100.0	Stains show:	Darks
		Opacity:	Translucent
		Color	
		/oxidation:	Dark green
		/reduction:	Lustrous multi-browns
		Note:	Some cracks

R 125 Pea Green

Colemanite	55	Temperature:	C/08–04
Nepheline syenite	20	Surface	
Plastic vetrox	15	@ C/06:	Gloss
Potassium dichromate	7	Fluidity:	Little
Tin oxide	2	Stains show:	None
Lithium	1	Opacity:	Opaque
	100	Color	
		/oxidation:	Pea green
		/reduction:	Dark pea green

R 126 Jade Green

Colemanite	50.0	Temperature:	C/08–06
Plastic vetrox	20.0	Surface	
Barium		@ C/08:	Gloss
carbonate	10.0	Fluidity:	Some
Lithium	10.0	Stains show:	Darks
Nepheline		Opacity:	Translucent
syenite	8.1	Color	
Chrome oxide	1.6	/oxidation:	Jade green
Cobalt		/reduction:	Dark jade
carbonate	.3		green
	100.0	Note:	Cracks

R 127 Grass Green

Colemanite	48	Temperature:	C/010–08
Borax	42	Surface	
Red copper		@ C/010:	Gloss
oxide	6	Fluidity:	Fluid
Rutile	4	Stains show:	Yes
	100	Opacity:	Transparent
		Color	
		/oxidation:	Grass green
		/reduction:	Opaque
			copper
			luster
		Note:	Cracks

R 128 Leadless Green

Colemanite	44	Temperature:	C/013–010
Borax	43	Surface	
Red iron		@ C/013:	Gloss
oxide	8	Fluidity:	Fluid
Black copper		Stains show:	Darks
oxide	5	Opacity:	Translucent
	100	Color	
		/oxidation:	Dark green,
			marbled
		/reduction:	Opaque,
			copper
			luster
		Note:	Some cracks

R 129 Leadless Tan

Colemanite	62	Temperature:	C/08
Nepheline		Surface	
syenite	20	@ C/08:	Semi-gloss
Plastic vetrox	10	Fluidity:	Some
Zircopax	4	Stains show:	Yes
Red iron		Opacity:	Translucent
oxide	4	Color	
	100	/oxidation:	Tan
		/reduction:	Tan
		Note:	Cracks

R 130 Tan

Colemanite	60	Temperature:	C/09
Frit #14		Surface	
(leadless)	30	@ C/09:	Gloss
Zircopax	5	Fluidity:	Little
Rutile	5	Stains show:	Yes
	100	Opacity:	Translucent
		Color	
		/oxidation:	Warm tan
		/reduction:	Cool tan
		Note:	Cracks

R 131 Light Tan

Colemanite	50	Temperature:	C/05
Red clay	20	Surface	
Soda ash	20	@ C/05:	Gloss
Nepheline		Fluidity:	Some
syenite	10	Stains show:	Yes
	100	Opacity:	Transparent
		Color	
		/oxidation:	Light tan
		/reduction:	Grayed tan
		Note:	Cracks

"Raku Form with Feathers," Tom Marsh;
Raku clay, beads, feathers, Raku lithium blue glaze;
15-1/4" H.

R 132 Yellow-Tan

Colemanite	30	Temperature:	C/08
Frit #14		Surface	
(leadless)	30	@ C/08:	Semi-gloss
Plastic vetrox	30	Fluidity:	Some
Yellow ochre	8	Stains show:	Darks
Red iron		Opacity:	Translucent
oxide	2	Color	
	100	/oxidation:	Yellow-tan
		/reduction:	Grayed tan
		Note:	Some cracks

R 133 Red-Brown

Colemanite	48	Temperature:	C/08
Plastic vetrox	20	Surface	
Boric acid	14	@ C/08:	Gloss
Manganese		Fluidity:	Some
carbonate	10	Stains show:	Yes
Red stain	8	Opacity:	Transparent
	100	Color	
		/oxidation:	Red-brown
		/reduction:	Do not reduce
		Note:	Cracks

R 134 Medium Brown

Colemanite	50.6	Temperature:	C/08
Cullet	26.2	Surface	
Plastic vetrox	9.1	@ C/08:	Semi-gloss
Zircopax	8.1	Fluidity:	Some
Red iron		Stains show:	None
oxide	4.0	Opacity:	Opaque
Manganese		Color	
carbonate	2.0	/oxidation:	Medium brown
	100.0	/reduction:	Medium brown
		Note:	Mix well; apply evenly with a thin coat

Glazing

The application of glaze is an involved part of the Raku process. The characteristics of the individual piece determine which glaze application method to use: dip, pour, spray, brush, or spatter. Raku forms require a thicker glaze than that used for typical stoneware or porcelain. Too thin a

"Raku Jar," Don Schaumburg; Raku, white glaze and copper oxide; 14" H × 12" W.

glaze will give an uneven, rough texture, spoiling the overall appearance. A uniform layer of glaze produces a smooth, even quality that is pleasant to touch. To help hold the glaze to the clay form, add a binding agent like CMC, gum tragacanth, or gum arabic. This will prevent the glaze from flaking or peeling off during the drying and the early part of the firing. Some glazes, after drying on the ware, are dusty; the glue will help keep them from coming off during handling. Also, the glue will keep the ingredients in suspension in the glaze bucket and prevent them from settling.

The techniques of overglaze, underglaze, stain wash, multiple glazes, engobes, silk-screen design, painted design, and so on can be used to decorate the ware. If you plan carefully, you can combine in one firing several glazes having the same maturing temperature and decoration techniques.

Figure 3.2

Applying glaze to bisque-fired form.

Firing

The most captivating part of the Raku process is the rapid firing, reducing, and rapid cooling. After you apply the glaze(s) (fig. 3.2), you can sun-dry the ceramic form or place it on top of the hot kiln to dry and preheat (fig. 3.3). Using heat from electricity, fuel oil, natural gas, open pit

Figure 3.3

Raku kiln with forms drying on top.

Figure 3.4

Planter set just inside kiln door to dry completely.

wood, diesel fuel, propane, or charcoal, heat the kiln to the desired temperature. Remove the thoroughly dried form from the top of the kiln and place it just inside the door of the kiln (fig. 3.4), using asbestos gloves and long metal tongs. The door of the kiln should be left open about four inches to allow the form to heat slowly. Then re-open the door and quickly place the form all the way inside the kiln, leaving enough space between forms for even glaze maturity and for the tongs (fig. 3.5). If the form is thoroughly dry and preheated, the clay should not explode or crack. Some glazes will bubble and boil for about ten minutes, then smooth out to cover the surface evenly.

Firing time ranges from 15 to 45 minutes depending upon the fuel, kiln variables, and the size of the form. Check the surface of the form through the peep-hole; when it is smooth and even, open the kiln and remove the ceramic form with the tongs (fig. 3.6). The tongs will leave slight marks on the soft glaze surface, one of the characteristics of Raku ware. The spectacular part of Raku firing occurs when the red-hot piece is removed from the kiln, cooled rapidly, then

Figure 3.5

Planter inside hot kiln.

Figure 3.6

Planter ready for removal from kiln.

handled as a new pot freshly glazed just minutes before.

OXIDATION COOLING

For oxidation cooling, place the hot ceramic form, directly from the kiln, on a brick or an asbestos pad to cool to air temperature or until the red glow disappears completely. Then pick it up (still using tongs) and place it in a bucket of water to cool to finger-handling temperature. The glazes and clay will have typical oxidation colors.

REDUCTION COOLING

To cool by reduction, remove the red-hot form from the kiln and immediately place it in a metal container filled with organic material such as grass, rags, sawdust, fine wood chips, crumpled newspaper, straw, dry leaves, or anything else that will burn (fig. 3.7). Quickly close the container with a tightly fitting fireproof lid. The hot form will ignite

Figure 3.7

Planter in metal container filled with leaves.

the organic material. The insufficient oxygen will cause the material to generate a great deal of smoke, creating a strong reducing atmosphere to cause color and chemical changes in the clay and glazes.

Glazes containing silver, copper, bismuth, and other metallic oxides may develop luster or pearlish surfaces. Carbon penetrates the pores of the clay, causing the clay to turn black or charcoal gray. Several glazes will turn one color under oxidation or neutral atmosphere firing, another under reduction. Some metallic oxides and carbonates will give up their oxygen molecules under reduction conditions, reverting to a pure metallic state and depositing a thin metal film on the surface of the glaze; these metallic surfaces will tarnish and take a polish like any other metal. Many luster glazes can be used for the Raku process. Some high-lead-content glazes may bubble or blister in the heavy reduction atmosphere.

Copper and other oxides in some glazes will develop a luster or metallic film in a heavy reduction atmosphere. Precooling will help avoid the luster and numerous glaze marks. Remove the form from the kiln and hold it with tongs for one or two minutes in the open air. Then place it in the metal container. As the form cools slightly, the glaze stiffens, resulting in fewer blisters and marks on the glaze, and less metallic film.

The length of time you let the form reduce and cool in the metal container depends upon the type of glaze, the size of the clay form, and the amount and type of burnable material. The more combustible the material, the faster the reduction. You can create texture and patterns on the glaze surface by placing leaves, wood chips, burlap, or the like on or under the piece when you set it in the metal container. The tong marks, ash marks, variegated surfaces, and colors will be different for each firing. The unpredictability of the outcome gives Raku its charm and mystique. (Figure 3.8 is the finished planter.)

If you remove the ceramic form from the metal container too soon, the re-oxidation of the glaze will cause the disappearance of lusters, red colors, and some multicolor effects. The metal container must have a tight seal to keep oxygen from reaching the glaze. Keep a small form in the container for about 20 minutes; larger pieces will take longer. After you remove it from the container, submerge it in water to quick-cool or set it on bricks or on an asbestos pad to cool to finger-handling temperature. The carbon, tar, pitch, and ash deposited on the glaze surface by the burning materials in the container can be removed by scrubbing the piece lightly with sand or other cleaning agents.

Figure 3.8

"Footed Planter," Randy Schneider;
Raku glaze over white clay;
10-1/2 "H × 7-1/2" D.

"Wine Decanter with Stand," H. James Stewart;
stoneware, oxides, salt-fired; 16-1/4" H.

four

salt-glazing

The earliest known use of salt to create a glaze was during the twelfth century, in potteries in or near Cologne, in Bavaria, and in Flanders. Between 1744 and 1749 Thomas Wedgwood further improved and developed the salt-glaze at the Churchyard Work in England, achieving a considerable reputation. The techniques of salt-glazing were brought from Germany and England to America in the early eighteenth century. By the late eighteenth century several American potteries were manufacturing salt-glazed jugs, crocks, jars, storage vessels, butter churns, bottles, and mugs. (English porcelain still dominated the American market for tableware and decorative ware.) The early stoneware forms, brown without decoration, were later replaced by gray-colored ware with cobalt oxide line and brush decorations. Grayware with blue cobalt decorations became a trademark of American salt-glazed stoneware. The decorations included simple flower, animal, fish, and landscape designs as well as involved patterns. The shapes of containers changed from the ovoid forms with small bases produced during the eighteenth and early nineteenth centuries to more utilitarian forms with wider bases, flattened sides, and heavier rims, produced even today. Salt-glazed stoneware was originally used to pickle, salt, can, and store food and liquids. The need for salt-glazed stoneware was reduced by the introduction of low-priced glass vacuum canning jars and by improved transportation of food to the marketplace. The use of salt-glazed stoneware in households and on farms declined to the point that the manufacturing of salt-glazed stoneware was almost nonexistent by the early twentieth century.[1]

Today there is considerable interest in reviving salt-glazing as a ceramic technique, for both its nostalgic and its aesthetic qualities. Among its attractive features are

[1] Donald B. Webster, *Decorated Stoneware Pottery of North America* (Rutland, Vt.: Charles E. Tuttle Co., 1972), pp. 19–23.

economy and simplicity: an inventory of expensive raw materials is unnecessary, and the process eliminates the need to apply glazes and requires only a single firing.

A salt-glaze is a hard, transparent, vitreous coating formed on the surface of clay. A slight orange-peel texture is produced on the surface by introducing salt into the kiln at the peak of the firing. A chemical reaction takes place as the salt volatilizes into sodium vapors that mix with the silica in the clay, creating a thin coating of soda-silicate glaze on the clay surface.

Salt-glazing offers the excitement and challenge of the firing as well as a special appreciation for the clay and its surface. Heavily patterned textures and contours of the clay form all respond to the touch of the salt-glaze, which reveals them clearly. To do salt-glazing you will need experience in gas kiln firing, a kiln suitable for salting, and a familarity with the salting process and safety procedures.

Clay Bodies for Salt-Glazing

Commercial salt-glazed products such as crocks, coping tile, sewer pipe, and jugs have been produced for several hundred years using various clay bodies. While most types of clay bodies will take a salt-glaze, several will not, including low-temperature earthenware and Raku bodies. The silica content, the firing temperature for the salting, and the maturing temperature of the body will determine its salt-fusing ability. Clays with a high ratio of silica to alumina are usually most successful. White, gray, tan, or brown stoneware and porcelain clay bodies, when fired in a salt kiln, achieve many subtle, unusual effects not always produced by the typical oxidation or reduction firing. If a

very thin coating of glaze is applied, the fired glaze surface will appear immature. Textures on the surface, whether subtle or heavily carved, will pick up the thin salt-glazed coating, because the glaze tends to pool in low areas and to coat thinly on raised areas, emphasizing fine details.

Although salt-glazing can be done to most types of clays, there are several requirements for successful salting:

1. The clay body must be vitrified or nearly so at the time of salt-glazing; otherwise it may absorb the glaze.
2. The ratio of Al_2O_3 (alumina) to SiO_2 (silica) must fall within certain limits, from $1\ Al_2O_3 : 3.3\ SiO_2$ to $1\ Al_2O_3 : 12.6\ SiO_2$.
3. The content of iron and manganese in the clay body must be under 6 percent; otherwise the clay will bloat and the glaze will be blistered and rough.
4. The temperature must be high enough to volatilize the salt compound, producing a smooth, even glaze finish.

Clay bodies low in silica content can be improved to take a better glaze by adding flint or silica sand. Another approach is to spray a thin coating of silica or sodium aluminosilicate, mixed with glue, onto the greenware or bisqueware. Although some clay bodies have the proper ratio of alumina to silica, various other factors can influence the process, preventing good salting. Most stoneware clay bodies are composed of kaolin, ball clay, fireclay, flint, and feldspar. They can be improved by the addition of 5 to 10 percent flint or a small amount of ball clay, bentonite, or macaloid to increase plasticity, and calcined kaolin to decrease shrinkage.

Clays found in nature have many excellent

properties but may not always suffice as a salt clay body. Raw clays can be found in many places throughout the world, as outcroppings in fields or in earth exposed by highway cuts, erosion, or land development. Most clay deposits are too small for commercial use but can be dug and used for individual needs. The clay is usually covered by several inches to several feet of dirt, in addition to grass, fences, or rocks. Remove this overburden and dig down to the uncontaminated clay, which you can remove with a pick or shovel and dry in the sun. After several weeks, crush or hammer the raw clay to break the hardest lumps to a fine dust. Screen the ground clay to remove impurities and large lumps. Seldom will you find a natural clay good enough to be a salt body; usually you will need to blend several clays, silica, and fluxes to obtain the desired color, hardness, plasticity, and working characteristics. Weigh the individual clay according to the clay body formula; blend it with the other clays, silica, and fluxes; and add water. Place the mixture in a pugmill or filter press or on drying bats to bring the body to a desired working consistency. Although you must spend a great deal of time and effort to prepare clay from the ground to throwing consistency, the process will be rewarding.

Many clay bodies can be used for salting with few or no alterations. We gathered formulas for several other clay bodies by experimenting, researching in publications, and exchanging with other ceramists. Each formula was subjected to a uniform testing procedure that analyzed (1) the ability of the clay to hold the salt-glaze; (2) glaze retention on a smooth surface; (3) percentage of clay shrinkage; (4) plastic quality of the clay; (5) fired color of the clay; and (6) recommended firing temperature. Each formula was measured into 100-gram batches, and water

was added to moisten the clay to throwing consistency. Part of the moist clay was hand-rolled into a rope and reduced in thickness until it crumbled. Plasticity was rated on a scale of 0 to 10: 10 is very plastic, 8 is plastic, 6 is adequate for hand-building, 4 is short, 2 crumbles easily, and 0 is unusable. To determine the clay color, shrinkage, and glazing ability, the balance of the moist clay was formed into a rectangular bar 4 1/2″ long by 1″ wide by 3/8″ thick. Two lines were etched, 10 centimeters apart. A 1″-wide area was painted with a black stain. The bar was air-dried, placed in a kiln, fired to the suggested temperature, and salted. After the test piece cooled, it was measured to determine the percentage of shrinkage. The glaze was analyzed for transparency, color, cracks, bubbles, chemical reaction, and surface. Samples of salt-glazed clay body testings are shown in figure 4.1. The data for each formula tested are given in the accompanying chart (fig. 4.2).

Figure 4.1

Salt-glazed clay body testings.

Figure 4.2
Clay body formulas suitable for salt-glazing.
Explanation: *Temperature* refers to the recommended
temperature range for firing. *Shrinkage* is the percentage
of shrinkage from the moist stage to the salt-fired stage.
Glaze quality is the surface quality, texture, and transparency.
Plastic scale ranges from 10 (very plastic) to 0 (unusable).
Color is the clay color resulting from firing in a salt atmosphere.

SGB 1 EPK Salt Body

Edgar plastic kaolin	50	Temperature:	C/7–10
Kingman feldspar	15	Shrinkage @ C/9:	13%
Flint	15	Glaze quality:	Smooth, even orange peel
Plastic fire clay	15		
Silica sand	5	Plastic scale:	8
	100	Color:	Warm white

SGB 2 Kentucky Ball Salt Body

Kentucky ball #4	45	Temperature:	C/7–10
Plastic fire clay	30	Shrinkage @ C/10:	13%
Kingman feldspar	10	Glaze quality:	Even salt surface
Flint	5	Plastic scale:	8.5
Silica sand	5	Color:	Warm ochre
Kaolin	5		
	100		

SGB 3 Jordan Salt Body

Jordan clay	45	Temperature:	C/8–10
Plastic fire clay	25	Shrinkage @ C/10:	13%
Flint	10	Glaze quality:	Smooth gloss
Kingman feldspar	10	Plastic scale:	8.5
Barnard clay	10	Color:	Dark brown
	100		

SGB 4 Chocolate Brown Salt Body

Jordan clay	40	Temperature:	C/8–10
Plastic fire clay	30	Shrinkage @ C/10:	14%
Kentucky ball #4	15	Glaze quality:	Smooth, high gloss
Flint	6	Plastic scale:	8.5
Kingman feldspar	6	Color:	Rich chocolate brown
Red iron oxide	3		
	100		

SGB 5 Fire Clay Salt Body

Fire clay	34	Temperature:	C/8–10
Jordan clay	34	Shrinkage @ C/10:	12%
Kentucky ball clay	17	Glaze quality:	Smooth, even coating
Kingman feldspar	6		
Flint	6	Plastic scale:	8.5
Silica sand	3	Color:	Warm tan
	100		

SGB 6 Tennessee Clay Salt Body

Tennessee ball clay	30	Temperature:	C/7–10
Fire clay	20	Shrinkage @ C/10:	13%
Jordan clay	20	Glaze quality:	Semi-gloss
Sewer pipe clay	20	Plastic scale:	8
Grog (60 mesh)	9	Color:	Broken medium tan-brown
Bentonite	1		
	100		

"Cup, Stemmed Goblet, Cup," Thom Collins;
stoneware, stains, glazes, vapor-glazed;
goblet 7" H.

"Casserole," William C. Alexander;
stoneware, stain, salt-fired;
5-1/2" H × 6-1/2" diam.

SGB 7 Fire Clay Salt Body

Plastic fire clay	28	Temperature:	C/7–10
Kentucky ball clay	25	Shrinkage @ C/10:	12%
Local red clay	15	Glaze quality:	Smooth gloss
Plastic vetrox	15	Plastic scale:	8.5
Flint	8	Color:	Rich dark brown
Grog (60 mesh)	7		
Red iron oxide	2		
	100		

SGB 8 Warm Tan Salt Body

Edgar plastic kaolin	26.3	Temperature:	C/0–10
Plastic fire clay	26.3	Shrinkage @ C/10:	12%
Kentucky ball #4	26.3	Glaze quality:	Slight sand finish
Plastic vetrox	15.8	Plastic scale:	8.5
Silica sand	5.3	Color:	Warm tan
	100.0		

SGB 9 Tan Salt Body

Jordan clay	25	Temperature:	C/8–10
Plastic fire clay	25	Shrinkage @ C/10:	13%
Kentucky ball #4	25	Glaze quality:	Smooth gloss
Plastic vetrox	15	Plastic scale:	9
Flint	10	Color:	Tan
	100		

SGB 10 Five Twenties Salt Body

Kaolin	20	Temperature:	C/9–10
Kentucky ball #4	20	Shrinkage @ C/10:	12%
Kingman feldspar	20	Glaze quality:	Semi-gloss
Flint	20	Plastic scale:	6
Plastic vetrox	20	Color:	Off white
	100		

SGB 11 Off White Salt Body

Potash feldspar	20	Temperature:	C/7–10
Plastic fire clay	20	Shrinkage @ C/9:	12%
Plastic kaolin	20	Glaze quality:	Semi-gloss
Kentucky ball #4	20	Plastic scale:	8
Flint	15	Color:	Off white
Grog (60 mesh)	5		
	100		

"Bean Pot and Pitcher," Karen Karnes;
stoneware, stains, natural with salt glaze;
pitcher 7-1/4" H.

Colored Clay

To color a clay body, add ceramic pigments to the clay body or apply engobes, stains, or oxides to the clay surface. You can achieve other surface changes by spattering or painting colemanite, feldspar, nepheline syenite, whiting, or frits. Further surface changes are caused in the kiln by flame pattern, character of the atmosphere, amount of load, chemicals in other clays in the kiln, placement of one object in relation to other objects, fuming, and the whim of the kiln gods.

COLORING CLAY BODIES

Most clay bodies are gray, tan, brown, or white. One of the oldest techniques for giving color to finished ceramics is to introduce color directly into the clay body, by mixing either colored clays or metallic oxides with the clay body. Natural clays such as albany slip, barnard clay, and even local red clays can be mixed with stoneware or porcelain clays for added color. The salt-glazing technique intensifies the clay color, and mixing colorants into the clay body will multiply the variety of colors and effects. You can achieve attractive results by decorating a basic clay form with colored-clay sprigging, appliqué, textures, relief, mishima,[2] and the like.

The following chart shows the results of introducing colorants into the clay body, then salt-glazing. For this testing, dry white clay was mixed with colorants, fired to C/9, and salted.

---

[2] *Mishima* is a Japanese term for a decoration technique. The surface of plastic clay is incised, then colored clays are placed in the incisions for a smooth surface. The form is then glazed and fired.

SALT-GLAZED COLORED CLAYS

Clay Colorant	Percentage	Resulting Color
Chrome oxide	4	Chrome green
Cobalt oxide	1	Light blue
Cobalt oxide	3	Dark blue
Copper oxide	4	Blue
Crocus martis	5	Rust
Ilmenite	4	Specks
Iron oxide, black	8	Brown
Iron oxide, red	3	Tan
Iron oxide, red	6	Brown
Manganese dioxide	5	Purple
Nickel oxide	4	Gray
Rutile	5	Tan
Umber, burnt	4	Tan

Commercial Stains	Percentage	Resulting Color
Black	8	Black
Blue	4	Blue
Brown	6	Brown
Green	4	Green
Orange	8	Orange
Tan	4	Tan
Yellow	8	Yellow

ENGOBE

One of the most important aspects of salt-glazing is that only one firing is necessary, since dry greenware can be placed in the kiln and fired. You can apply colored engobes and stains to wet, damp, or dry greenware to achieve a variety of colors. Engobes and stains can be applied as a wash, painted, sponged, spattered, or sprayed. Using engobes or stains to color the clay surface, instead of introducing color into the clay body, has several advantages: it reduces the inventory of different clay bodies that you need; it is an easy technique; and it offers a greater variety of colors. Of the many engobe formulas available, the easiest and often the most reliable engobe is made by mixing colorants with the clay body you are using. Formulas for making engobes can be found in the books listed in the bibliography. A good engobe formula for testing engobe colorants follows:

Flint	40
Fire clay	20
Calcined clay	20
Potash feldspar	10
Zircopax	5
Borax	5
CMC (glue)	1
	101

We conducted a testing series to determine the range of colors possible using a single engobe formula. We weighed the formula given above, added oxides, mixed it with water, and painted it on both porcelain and stoneware dry greenware tiles. We placed the tiles in a gas-fired kiln, fired the kiln to Cone/9, reduced slightly, and salted. Figure 4.3 shows sample engobe testings. The colorant, percentage used, and resulting colors over both porcelain and stoneware clays are given in figure 4.4.

Figure 4.3

Sample colored engobe testings.

Figure 4.4
Salt-Glazed Colored Engobe Testings

Oxide	Percent	Resulting Colors	
		Brown Stoneware	Porcelain
Cerium oxide	6.0	Slight ivory	Slight ivory
Burnt umber	6.0	Brownish	Brownish
Crocus martis	4.0	Brownish	Brownish
Chrome oxide	6.0	Chrome green	Chrome green
Cobalt oxide	4.0	Dark cobalt blue	Dark cobalt blue
Cobalt oxide	2.0	Medium cobalt blue	Medium cobalt blue
Copper oxide	2.0	Slight pink	Slight pink
Ilmenite	6.0	Warm specks	Warm specks
Iron oxide, black	6.0	Dark brown	Dark brown
Iron oxide, black	2.0	Slight tan	Slight tan
Iron oxide, black	8.0 ⎫		
Manganese dioxide	3.2 ⎬	Black, semi-gloss	Green-black
Cobalt oxide	1.2 ⎭		
Iron chromate	4.0	Medium gray	Medium gray
Iron oxide, red	8.0	Dark brown	Dark brown
Iron oxide, red	4.0	Tan	Tan
Manganese dioxide	8.0	Dark red-purple	Dark red-purple
Manganese dioxide	4.0	Light red-purple	Light red-purple
Nickel oxide	6.0	Gray brown	Medium gray
Nickel oxide	2.0	Brown	Gray brown
Rutile	8.0	Dark ivory	Dark ivory
Rutile	2.0	Medium ivory	Medium ivory
Tin oxide	10.0	Cool white	Cool white
Yellow ochre	6.0	Warm tan	Warm tan
Zircopax	10.0	Warm white	Warm white

STAINS

Like engobes, stains can be applied to wet, damp, and dry greenware as well as to bisqueware. Most commercially and individually made stains will decorate the clay surface satisfactorily, but some stains will not provide a sufficient silica surface to allow the vaporized salt to form a salt-glaze. An overly thick coating of stain will cause bubbles, craters, or running of the glaze. Reds, oranges, and some yellow stains will burn out or produce gray colors at high temperatures.

Some techniques for using stains and engobes to decorate clay surfaces for salting are illustrated in figure 4.5. (1) The entire surface of stein D, including any texture, relief, carved design, or stamped pattern, was painted with the stain; then the excess was rubbed off, allowing the stain to remain in the recessed areas for contrasts and color. (2) The stain was painted on the background areas around the designs, patterns, relief, or textures of stein C. (3) On stein B, stain was applied in the recessed areas of incised drawings or designs; then any excess was removed, allowing the stain to remain only in the recessed areas. Stein A is natural clay.

Figure 4.5

Salt-glazed German steins.

"Salt and Pepper Shakers," Milt Beens; stoneware bodies, salt-glazed; 4-1/4" to 5" H.

ARTIFICIAL SALT-GLAZING

Artificial salt-glazing involves coating the clay form with a salt compound (ground salt, flint, and soda) that produces a salt-glaze when fired. The soda, salt, and silica combine at C/4 to C/9 to form a glaze. During the firing, an envelope of glaze-forming fumes may develop a blush of glaze on any nearby clay form. The following formula is the basis for developing an artificial salt-glaze:

Artificial Salt-Glaze

Soda ash	40
Ground salt	30
Flint	25
Bentonite	3
Glue (CMC)	2
	100

Temperature:	C/4–10
Surface @ C/6:	Slight gloss
Glaze flow:	Little
Transparency:	Clear
Color:	Clear

OXIDES UNDER THE SALT-GLAZE

A quick and easy method of introducing color and texture to the surface of clay is to paint, sponge, spray, spatter, or wipe a metallic oxide onto the surface of the clay form. Painting on the oxide and then wiping it off create an antique effect and bring out the details of highly decorative surfaces in the salting (see figure 4.5, stein C). To make an oxide wash, simply mix water, 10 to 20 percent flint, and the oxide. Figure 4.6 shows the colors resulting from oxides used on dark brown and porcelain clays which are then salt-glazed.

Firing

Salt-glazing is still widely used in industry to create an inexpensive glaze on a variety of ceramics, such as sewer pipe, hollow building blocks, roof and coping tile, conduit, silo blocks, beer mugs, crocks, dinnerware, storage jars, and vinegar jars. Many ceramists who are interested in salt-glazing use commercial techniques and

"Lidded Soup Tureen with Ladle," Robert Winokur; stoneware, stains, salted, fumed; 11-1/2" H × 8-3/4" W.

Figure 4.6
Colors of Oxides under Salt-Glaze

Oxide	Resulting Colors	
	Brown Stoneware	Porcelain
Albany slip	Dark chocolate brown	Dark chocolate brown
Barnard clay	Semi-matt charcoal brown	Semi-matt charcoal brown
Burnt sienna	Dark tan	Matt warm tan
Cerium oxide	Medium tan	Warm tan
Chrome oxide	Dark brown	Dark brown
Cobalt oxide	Matt black	Semi-matt; dark blue where thin
Copper oxide	Dark gray-brown	Brown
Copper phosphate	Dark brown	Matt browns
Crocus martis	Matt dark brown	Gloss-thin; matt-thick; medium brown
Ilmenite	Black specks	Dark brown specks
Iron chromate	Black	Black
Iron oxide, black	Medium brown	Medium brown
Iron oxide, red	Medium brown	Medium brown
Manganese dioxide	Semi-matt dark brown	Brown black
Nickel oxide	Dry matt black	Dry matt black
Rutile	Brown	Medium tan
Tin oxide	Tannish	Ivory
Titanium oxide	Dark brown	Matt black
Umber, burnt	Brown-black	Brown
Uranium	Brown	Yellow
Vanadium	Rich brown	Black
Yellow ochre	Matt broken tans	Matt tan
Zinc	Broken brown	Broken tan
Zircopax	Matt dark tan	Matt warm tan

facilities to obtain a wider range of both functional and sculptural ceramic forms. For example, Jerry Caplan of Pittsburgh, Pennsylvania, creates sculpture forms from sewer pipe clay direct from the pipe extruder. He has arranged with the industrial firm to place his sculpture in the kiln along with the sewer pipe for simultaneous salt-firing.

Salt-glazing uses many of the tools, procedures, techniques, and equipment that are used for regular stoneware and porcelain. Wood, coal, oil, propane, or natural gas can be used to heat the kiln. The firing is a simple process, but it requires your total involvement. The method of firing, the salting process, the kiln, and the shape of the pot depend upon the results you desire.

PROCEDURE

Ceramic forms can be made on the wheel or by hand using traditional stoneware or porcelain techniques. The surface can be enriched with textures, relief, designs, or other decorative methods. Engobes or stains are brushed, sprayed, dipped, spattered, or poured onto the surface, and the interior of closed forms is glazed. After drying thoroughly the forms are stacked in the kiln. (Some ceramists have successfully placed wet greenware in the kiln, turned on the burners, and closed the dampers, allowing the steam to create an envelope around the wet clay that keeps the temperature uniform, prevents hot spots, and allows the clay to dry evenly.)

KILN WASH

To prevent salt-glazed forms from adhering to the kiln furniture, coat the inside of the kiln, the shelves, and the kiln furniture with kiln wash. Before each firing, rub off excess wash and glaze drips before applying additional wash. Give the shelves two thin coats of kiln wash and allow them to dry before stacking them in the kiln (fig. 4.7).

Figure 4.7

Silicon carbide shelves coated thinly with high-alumina wash.

Commercial kiln washes are not recommended for salt-glazing because many of them contain silica. Individually formulated washes such as those listed below are cheaper and very satisfactory.

KW-1 Salt-Glaze Kiln Wash

Alumina oxide (or hydrate)	80
Sagger clay	18
Glue (CMC) or ball clay	2
	100

KW-2 Salt-Glaze Kiln Wash

Alumina oxide	50
Calcined kaolin	50
	100

KW-3 Salt-Glaze Kiln Wash

Zircon	60
Calcined kaolin	35
Ball clay	5
	100

Figure 4.8

Stacking the salt kiln.

Figure 4.9

Loading the salt kiln with ware to be re-fired.

KILN LOADING

In loading ware into the kiln, be sure to allow for good air circulation and to avoid having pieces touch each other (fig. 4.8). A good rule is to leave a finger thickness between each item and the shelves, walls, posts, and other items. Although some clays shrink as much as 10 percent during salt-firing, it is nevertheless wise to maintain the distance. Controlled stacking will allow flame lickings, salt build-up, and thick–thin glaze surfaces. To obtain controlled irregular salting, you can stack several pots on top of each other (with or without clay wads), group pieces close together, place bricks or shards next to the ware, lay the pieces on their sides, overlap pieces, or position the ware so that part of it extends over the edges of the kiln shelf. Load previously fired pieces (re-fires) in the center, away from the hot flames, so that they heat slowly (fig. 4.9).

Use clay wads to lift the ware off the kiln shelves and the lids off the pots, and to stack forms on top of each other. Remove a pinch from a piece of clay and roll it into a ball. Some salters roll the ball lightly in a pan of alumina oxide or kaolin; this dry material helps prevent the clay ball from sticking to the shelves and pots.

Make four to six *draw rings* from the same clay as that being salt-fired. Placed in the kiln (fig. 4.10), they can be removed easily to determine the amount of salt-glaze build-up.

Figure 4.10

Draw rings in place.

Figure 4.11

Bricking up the kiln with 2 layers of firebrick.

Also set in place the pyrometric cones, to help indicate when to start the salting. The cones are not always reliable if the kiln interior is heavily covered with salt-glaze; the glaze will revaporize during the heating and coat the cones. Close the door of the kiln, or brick up the opening (fig. 4.11); now the kiln is ready for preheating.

FIRING PROCEDURE

Turn on the burners to preheat the kiln and dry out the ware by bringing the temperature slowly to 500°F. At approximately 1500°, medium-reduce the kiln for 20 minutes or maintain a slight reduction atmosphere from 1500° to the salting temperature. The reduction action of the kiln atmosphere converts the ferric iron compounds in the clay, a process called *body reduction*. At C/8 to C/10 temperatures the iron helps to make a denser body, a darker color, and a smoother, glossier, less porous surface.

Various happy accidents can be achieved by placing organic matter such as apples, oranges, bananas, straw, or wood on the ware.

During the firing, the ash deposits mix with the sodium and silica to form a glazed area different in texture, color, and surface from the salt-glaze. For another ash variation, introduce wood directly into the kiln. When it lands on the ware, the ash produces wood-ash glaze areas, usually on the shoulders or high spots, and mixes with the salt-glaze. In a non–wood-burning kiln, thin strips of wood or bundled straw can be pushed into the fire box.

SALTING

During the last 30 degrees of firing, the damper is partially closed and common rock or table salt, slightly dampened with water, is introduced into the kiln. The water turns to steam, which breaks down the salt particles for faster melting, improving the glaze quality. *Note: Do not overly wet the salt or introduce large amounts of water into the kiln.* The steam created at high temperature can break brick, shelves, posts, and pots, and even blow up the kiln.

The salting is done in four to six batches, rather than in one, to produce a uniform coating. There are several methods of introducing salt into the kiln: it can be shoveled, packed in paper cups or paper bags (fig. 4.12), or blown in with a fan. In the drip

Figure 4.12

Pushing a paper bag filled with dampened salt into the fire box.

method salt water (brine) is dripped in front of the burners. The intense heat turns the salt into a liquid, which then volatilizes.

The clay body should be as vitrified as possible at the temperature fired for most suitable salt-glaze. The amount of salt necessary to deposit a desired glaze thickness depends upon air pressure, temperature, type of fuel, fullness of the kiln load, amount of salt placed in the kiln, silica content of the clay, types of burners, dampness of the salt, and type of kiln brick. Therefore, the amount and number of saltings vary. After several saltings, remove a draw ring from the kiln, immerse it in water, and inspect it. If it is underglazed, do several more saltings; remove more draw rings until the desired glaze thickness is reached. Holding the kiln at the peak temperature (soaking) will improve the quality of the glaze surface by healing any rough surface. Next, open the dampers and adjust the blowers for an oxidation firing to soak the glaze and convert the gray-colored ferrous silicate glaze to the warmer reddish-brown ferric glaze. Turn off the kiln, close the dampers, and allow the kiln to cool to room temperature. Slow cooling produces a darker clay body, while faster cooling (dampers open full and spyhole plugs removed) produces a lighter color.

"Vase," Don Reitz;
stoneware, sprayed oxides, salt-fired;
16-3/4" H × 10" diam.

OXIDATION FIRING

Stoneware develops cleaner colors (with less metallic spotting), lighter colors, and fewer blisters under oxidizing conditions. Some potters obtain better results using oxidation firing up to the last salting, then reduction. When a clay body contains a heavy amount of iron or manganese oxides, oxidation firing is necessary up to the last salting; otherwise a bubbly, spongy glaze will form over a bloated clay.

REDUCTION FIRING

The method of firing and salting depends upon the clay body. Clay that contains soluble salt tends to produce rough glazes when salted at 1700°F. The reduction atmosphere will generally decompose the calcium sulfate film (calcium is found in some salt compounds) and facilitate a better glaze. Don Reitz of Madison, Wisconsin, suggests firing the kiln to C/8 in oxidation, then reducing moderately to C/9, and finally salting. The amount of iron in the clay body is in direct proportion to the glaze thickness; the higher the iron content in the clay, the less salt is necessary.

UNLOADING THE KILN

When the temperature of the kiln reaches 500°F., remove the spyhole plugs and slightly open the door. (In the case of hand-bricked doors, remove the outer layer of bricks.) As the kiln cools, inch the door open accordingly; for the hand-bricked door, remove the top bricks of the inner layer. At approximately 300° you can completely open the door (fig. 4.13). Remove the ware when it is cool enough to handle (fig. 4.14).

SAFETY AND POLLUTION

Whether the kiln used for salting is indoors or outdoors, you must take several precautions: (1) ventilate the areas around the kiln;

Figure 4.14

"Punch Bowl," Stephen Nelson; salt-glazed stoneware with oxides, engobe, and glazes; 17-1/2" diam. × 10-3/4" H.

Figure 4.13

Kiln load before and after salt-firing.

(2) have a fume collection hood over the kiln to draw off flames, fumes, and smoke; and (3) make sure there is a tall stack to draw and disperse the dangerous acid fumes into the atmosphere and prevent them from frosting paint or metal or killing plant life. About 10 percent of the sodium in the salt introduced into the kiln comes out the stack, and the remaining 90 percent forms a glaze. The carbon dioxide and nitrogen gases create no particular problem; only the water-hydrogen-chlorine mixture is of concern. The sodium attracts chlorine and surrounds itself with moisture to produce a noticeable fog blanket that disperses in 15 to 30 minutes. Widely dispersed, the vapors are safe; concentrated, however, they are highly toxic.

Several commercial potteries and individual ceramists are experimenting with ways to eliminate or considerably reduce the amount of pollution from chlorine gas and hydro-chloric acid. One method is to introduce sprayed water (scrubbers) into a specially built stack as a means of collecting the

pollution. The drawback is that you must dispose of the acid water produced by this method. A second method is to build a large stack of insulating firebrick, with a maze-like interior. As the fumes travel through the labyrinth, they soak into the pores of the brick. Unfortunately, after a while the bricks become saturated and ineffective. A third method of reducing pollution is to line the kiln with high-alumina brick; less salt will be needed to salt the ware, but the cost of the brick is high. Still another method is to substitute a boron and soda compound for salt to create the vapor glaze; the gas released by this compound is not dangerous.

Salt

Most salt kilns use approximately one pound per cubic foot. A high-alumina brick kiln may take only a third of a pound, while a new, regular brick kiln may take one and a half pounds per cubic foot. Highway salt, rock salt, or table salt work satisfactorily; rock salt has the advantages of lower price, small grains, and ready availability. Too little salt results in thin or dull glazes; too much may cause the glaze to run. Powdered coal, charcoal dust, oil, sawdust, or wood shavings mixed with the salt and placed in the kiln will cause combustion. The more the air moves inside the kiln, the more even the glaze coating. Very rapid cooling or too low a salting temperature will cause a film on the glaze.

ADDITIVES TO SALT

Instead of using 100 percent salt for salting, try mixing up to 15 percent soda ash, borax, boric acid, and/or sodium bicarbonate with the salt without changing the salting process. The result will be a smooth, even salt-glaze that can be fired at a temperature lower by

"Plate," Don Pilcher; porcelain, salt over salt stain; 17-1/2" diam.

one or two cones. The following formula is a basis for experimentation:

Salt	85
Borax	10
Soda ash	5
	100

You can improve the glaze quality by adding other fluxes: for example, 2 percent zinc chloride, 2 percent lithium chloride, 10 percent borax, 8 percent boric acid, 8 percent sodium bicarbonate, 8 percent sodium chloride, or 5 percent potassium chloride. Introduce the salt-and-flux mixture into the kiln in five batches to give a continuous flow of fumes. An excellent method of salting with additives, developed at New York State College of Ceramics, improved all the glaze properties—thickness, smoother surface, better color, less crazing (network of glaze surface cracks), increased strength. For the first three saltings, salt is used alone; for the fourth salting a mixture of salt and 4 to 15 percent borax and boric acid is used; and for the final salting borax and boric acid alone are used.

COLORANTS MIXED WITH SALT

Color in the salt-glaze is achieved by mixing 0.0025 to 0.04 percent of carbonates, chlorides, or sulfates of metals with salt. Even these small amounts add color to all glaze surfaces. They may have a slight residual effect on the next firing, since some of the colorants may not burn out completely or may adhere to the kiln interior. Tom Turner of Liberty, South Carolina, achieves brilliant reds, soft pinks, and subtle red-purples on white clay using a high-alumina brick kiln. White and off-white clays bring out the subtle colors of the glazes. Metallic oxides in dark-colored clays will overpower the color subtleties.

Clay, kiln, burner system, and firing temperature directly influence color and glaze results. Figure 4.15 is a guide to colors resulting from specific percentages of colorants.

FUMING

Stannous chloride and other chemicals produce a mother-of-pearl or pearlescent effect on ceramics. Small batches of the compound are put into the kiln when it has cooled to a dull red heat. The burners or blowers are turned on and the dampers opened slightly to achieve continuous air movement. The fumes circulate around the kiln and deposit a thin coating. When the draw rings show the desired pearlescence, the dampers are closed, the burners are turned off, and the kiln is cooled to room

Figure 4.15
Colorants Mixed with Salt

Colorant	Percentage		Resulting Color
Cobalt carbonate	.0025	to .02	Bluish
Cobalt nitrate	.0025	to .02	Bluish
Cobalt sulfate	.0025	to .02	Bluish
Copper carbonate	.005	to .02	Blue to green to red
Copper nitrate	.05	to .03	Blue to green to red
Copper nitrate		.05	Reddish
Tin oxide		.05	
Iron sulfate	.01	to .03	Greenish
Manganese chloride	.01	to .03	Purplish
Manganese carbonate	.01	to .02	Purplish

temperature. The pearlescent effect is very attractive on salt-glazed, orange-peel textured, light-colored clay bodies. For details on fuming formulas and techniques, see chapter 5.

Non-Salt Salt-Glazing

Several communities have enacted ordinances forbidding or limiting salt-glazing because of the resulting sodium chloride and hydrochloric acid air pollution. Recent research provides an alternative to closing down the salt kiln or moving to a community with less stringent pollution laws. Sodium carbonate (soda ash), sodium bicarbonate (baking soda), borax, boric acid, sodium nitrate, and similar boron and soda-based chemicals alone or in combination produce glaze results very similar to those of salt-glaze and just as economical.

With the same techniques used for salt-glazing, the salt substitute is vaporized in the kiln's flames, causing the decomposition of the chemicals. The steam and sodium vapors combine with the silica of the clay to form a thin coating of glaze. Water and carbon dioxide, the by-products from the use of the alternative chemicals, are no more damaging than the by-products of regular stoneware or porcelain gas firing.

The amount of chemicals necessary to produce a smooth orange-peel glaze throughout the kiln is about three ounces per cubic foot, as compared to one pound of salt per cubic foot. A 50–50 mixture of the chemicals seems to work satisfactorily. Because the amount of chemicals in the kiln is low, a more active fire or more air movement is necessary. To obtain the air movement necessary to force the damp mixture into the kiln, many ceramists use innovative techniques such as sand blaster, heavy-duty blower, or drip feed in front of the burners.

"Neck Piece with Beads," Sylvia Hyman; stoneware, salt glaze over rutile slip; 9" × 10-1/2".

METAL VAPOR

Some metals, such as zinc and bismuth, produce surfaces similar to those produced by common salt. At the peak temperature, zinc (flakes or dust) is put into the kiln to volatilize and combine with the clay to form a glaze. The resulting colors are yellow, green, tan, or brown, depending upon the lime, silica, and colorants in the clay body. Bismuth subnitrate produces gray colors.

Kiln

Most ceramists use a down-draft kiln for salt-glazing. Industrial salt-glazing is done in tunnel and continuous kilns, which have a uniform distribution of heat and fumes. The salt is dropped into an opening at the top of the kiln or pushed into a

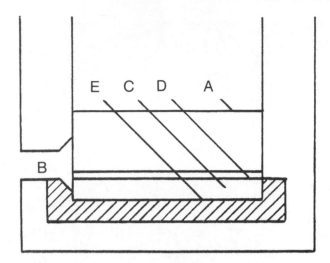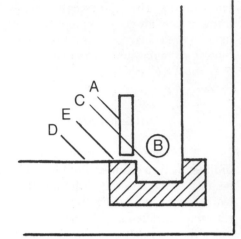

Figure 4.16

Side (left) and end (right) views of salt-kiln fire box:
(A) bag wall; (B) burner port; (C) fire box area;
(D) floor; (E) high-alumina brick or castable.

special fire box designed so that the force of the flames will spread the vapors. Dropping salt into the top of the kiln has the advantage that the salt vaporizes very quickly.

The fire box is an important part of the salt-firing kiln. The hottest part of the kiln, with the greatest potential for salt build-up, the fire box is where the salt is volatilized. If the fire box is built improperly, the salt-glaze may build up and erode under the bag wall, melting the pots to the floor and running out the burner ports onto the burners. These problems can be avoided with a recessed fire box made of high-alumina firebrick or high-alumina castable (figure 4.16).

HARD BRICK AND INSULATING BRICK KILNS

Insulating brick for salt-glazing has a short life because of the build-up of glaze deposits on the brick. Hard brick (refractory) lasts longer, but it will slowly deteriorate from the fumes and glaze build-up. Mortaring the brick with a thin coating of high-

alumina mortar is recommended; otherwise the glaze build-up will shift the brick.

Once the kiln is built, you should set up a firing schedule and keep firing records. Since the interior of the kiln builds up a salt coating, each firing in a new kiln will give slightly different effects until the kiln has mellowed and ripened. The character of the kiln is determined by salt build-up and other factors such as (1) flame size and deflectors established; (2) uniform temperature stabilized throughout the kiln; (3) dampers marked for kiln pressure; (4) hot spots eliminated; (5) amount of salt for firing established; (6) valve settings recorded; and (7) length of firing time established.

ALUMINA BRICK

In addition to the variables of stoneware, porcelain, and earthenware firing, the salt-glazing process adds variables of brick lining, atmospheric pressure, and the amount and type of salt. A major variable is the kiln's brick lining, which can have several compositions. High-alumina brick is desirable; it does not build up a glaze coating, it lasts much longer than other brick, and it will not flake, chip, or decompose like regular brick. It eliminates glaze drippings from the top of the kiln onto pots and shelves, which may or may not be an advantage; some ceramists like the happy accidents of unusual glaze drippings, flame markings, irregular glazing, and textures on the ware. The major disadvantage of high-alumina brick is its cost. At this writing, standard hard bricks cost $.50 each; standard 2600°F. insulating bricks cost $.90 each; 60 percent alumina bricks are $1.25 each; 70 percent alumina bricks $1.40; and 80 percent alumina bricks $1.60. High-alumina brick is up to three times more expensive than regular hard brick. In the long run, however, it is well worth the initial expense compared to the time and cost of later rebuilding a regular hard-brick kiln and redoing pots that are unusable because of undesirable salt drips. Since high-almina kilns also use less salt, the overall costs of high-alumina and regular hard-brick kilns may eventually equalize.

KILN SHELVES

Alumina and silicon carbide shelves have the longest life. Fireclay shelves often warp and/or sag under the weight of the pots at high temperatures. All shelves should be given a thin coating of kiln wash to protect them and to facilitate the removal of drippings.

Glazes

The insides of closed forms such as vases, jars, pitchers, and bottles will not have a salt-glaze coating, because the salt vapors will not enter the closed areas. These areas should be coated with a glaze that will mature at the firing temperature of the salting.

Most stoneware or porcelain glazes applied to bisqueware mature satisfactorily in the salting process. Most non-slip glazes, applied to wet or dry greenware, flake off the surface because of the shrinkage differences between the clay body and the glaze and because the greenware does not absorb the glaze into its pores. Many ceramists fire their ceramics once and often use slip glazes because they adhere to the clay body and shrink at the same rate as the drying clay. The traditional firing and glazing sequence used by most ceramists for stoneware or porcelain firing is to make a form, bisque-fire it, apply a glaze, then high-fire it. In salting, many ceramists make a form, apply a glaze to the inside, then salt-fire the form. Slip glazes are used most often, for they adhere to the clay body, shrink at the same rate as the drying clay, and have a long glaze-maturing range with little or no crazing and pinholing.

SLIP GLAZES

Slip glazes are made largely from clays that contain iron and other mineral impurities. Low-temperature clays such as earthenware, and even ground low-temperature pyrometric cones, when mixed with fluxes will melt into a glaze. Slip glazes are generally used at stoneware and porcelain temperatures. Local low-temperature clays can be used as slip glaze. Of the several available commercially the most popular is Albany slip, from the area near Albany, New York. Albany slip melts at Cone 7 and higher to produce dark brown colors. Colorants, opacifiers, fluxes, refractories, and other minerals added to slip create a variety of surfaces and colors. Slip glazes are the basis for the old, well-known Chinese glazes called partridge feather, oil spot, and Temmoku. Tans, blues, greens, browns, and blacks are possible, with surfaces of matt, semi-matt, semi-gloss, and gloss.

Figure 4.17 lists the results of slip glaze testings. Each formula was painted on two porcelain test pieces. One was fired at C/9 reduction, the other salt-glazed at C/9.

Figure 4.17
Slip glazes for the salting process

			Reduction	Salted
SG 101 Broken Tan				
Albany slip	85	Temperature:	C/9–10	C/9–10
Rutile	10	Surface @ C/9:	Semi-matt	Semi-matt
Zinc	5	Fluidity:	Some	Some
	100	Stains show:	No	No
		Opacity:	Opaque	Opaque
		Color:	Broken tans, yellows, and browns	Broken tan and gray crystals
SG 102 Olive				
Albany slip	80	Temperature:	C/9–10	C/9–10
Cornwall stone	20	Surface @ C/9:	Gloss	Matt
	100	Fluidity:	Some	Some
		Stains show:	No	No
		Opacity:	Opaque	Opaque
		Color:	Oil spot, brown	Olive
SG 103 Charcoal				
Albany slip	80	Temperature:	C/9–11	C/9–11
Red iron oxide	20.	Surface @ C/9:	Matt	Semi-matt
	100	Fluidity:	No	No
		Stains show:	No	No
		Opacity:	Opaque	Opaque
		Color:	Dark purple-brown	Charcoal and broken ochres
		Note:	Some small crystals	
SG 104 Brown				
Albany slip	80	Temperature:	C/8–9	C/8–9
Nepheline syenite	15	Surface @ C/9:	Gloss	Gloss
Lithium	5	Fluidity:	Fluid	Fluid
	100	Stains show:	No	Yes
		Opacity:	Opaque	Transparent
Color:		Color:	Oil spot brown and blacks	Brown
SG 105 Yellow-Brown				
Albany slip	70	Temperature:	C/9–10	C/9–10
Burnt sienna	10	Surface @ C/9:	Semi-matt	Semi-matt
Cornwall stone	10	Fluidity:	Little	Little
Zircopax	10	Stains show:	No	No
	100	Opacity:	Opaque	Opaque
		Color:	Purple-brown	Broken yellow-brown
SG 106 Green-Tan				
Albany slip	66.7	Temperature:	C/9	C/9
Potash feldspar	16.7	Surface @ C/9:	Gloss	Matt
Dolomite	8.9	Fluidity:	Some	Some
China clay	7.7	Stains show:	Darks	Darks
	100.0	Opacity:	Translucent	Opaque
		Color:	Greenish brown	Medium green-tan

			Reduction	Salted

SG 107 Dark Green

			Reduction	Salted
Albany slip	60	Temperature:	C/9	C/9
Nepheline syenite	35	Surface @ C/9:	Gloss	Matt
Yellow ochre	5	Fluidity:	Some	Some
	100	Stains show:	No	No
		Opacity:	Opaque	Opaque
		Color:	Dark tans and browns	Green

SG 108 Steel Blue

			Reduction	Salted
Albany slip	55	Temperature:	C/9	C/9
Potash feldspar	40	Surface @ C/9:	Gloss	Gloss
Whiting	4	Fluidity:	Some	Some
Cobalt carbonate	1	Stains show:	Darks	Darks
	100	Opacity:	Translucent	Transparent
		Color:	Dark bluish	Steel blue
		Note:	Cracks	Cracks

SG 109 Light Tan

			Reduction	Salted
Albany slip	50	Temperature:	C/9	C/9
Cornwall stone	40	Surface @ C/9:	Gloss	Gloss
Whiting	5	Fluidity:	Some	Some
China clay	5	Stains show:	Yes	Yes
	100	Opacity:	Transparent	Transparent
		Color:	Greenish tan	Light tan
		Note:		Cracks

SG 110 Gray-Black

			Reduction	Salted
Albany slip	53.6	Temperature:	C/9	C/9
Cornwall stone	30.6	Surface @ C/9:	Gloss	Semi-gloss
Whiting	8.4	Fluidity:	Fluid	Fluid
Red iron oxide	7.4	Stains show:	No	No
	100.0	Opacity:	Opaque	Opaque
		Color:	Browns and blacks	Oil spotting; gray-black and olive

SG 111 Broken Tan

			Reduction	Salted
Albany slip	50	Temperature:	C/9–10	C/9–10
Potash feldspar	30	Surface @ C/9:	Gloss	Semi-gloss
Zircopax	20	Fluidity:	Some	Some
	100	Stains show:	No	No
		Opacity:	Opaque	Opaque
		Color:	Broken tan	Broken tan

STONEWARE GLAZES

Whenever you want the inside or other parts of a pot to have a lighter color than the color produced by slip glazes, you must use regular glazes. Most regular glazes will work satisfactorily in the salt-firing, but the clay form should first be bisque-fired. Then glaze is applied and fired with the salt-firing that can contain both bisque and greenware.

Testing is necessary because some glazes become very fluid under salting. We tested several formulas designed for salt-firing by painting each of them on two porcelain test pieces. One was salt-glazed at C/9, the other reduction-fired at C/9. The formulas and results are given in figure 4.18.

Figure 4.18

Stoneware glazes used with salt-glazes			Reduction	Salted

SSG 1 Cornwall Stone Glaze

			Reduction	Salted
Cornwall stone	78	Temperature:	C/9–10	C/9–10
Whiting	14	Surface @ C/9:	Gloss	Semi-matt
China clay	8	Fluidity:	Some	Some
	100	Stains show:	Yes	Yes
		Opacity:	Transparent	Translucent
		Color:	Clear	Milky
		Note:		Cracks

SSG 2 Kingman Glaze

			Reduction	Salted
Kingman feldspar	62	Temperature:	C/9	C/9
Dolomite	26	Surface @ C/9:	Gloss	Matt
China clay	12	Fluidity:	Some	Some
	100	Stains show:	Yes	Yes
		Opacity:	Transparent	Translucent
		Color:	Clear	Whitish
		Note:	Cracks	Cracks

SSG 3 Opaque Glaze

			Reduction	Salted
Kingman feldspar	57	Temperature:	C/9–10	C/9–10
Barium carbonate	22	Surface @ C/9:	Matt	Matt
Whiting	10	Fluidity:	Little	Little
Tin oxide	6	Stains show:	No	No
China clay	5	Opacity:	Opaque	Opaque
	100	Color:	Cool white	Cool white

SSG 4 Jack Troy's Clear Glaze

			Reduction	Salted
Kingman feldspar	55	Temperature:	C/9	C/9
Flint	18	Surface @ C/9:	Gloss	Gloss
Whiting	14	Fluidity:	Some	Some
Zinc oxide	9	Stains show:	Yes	Yes
Kaolin	4	Opacity:	Transparent	Transparent
	100	Color:	Clear	Clear
		Note:	Some cracks	Some cracks

SSG 5 Transparent Glaze

			Reduction	Salted
Kingman feldspar	50	Temperature:	C/9	C/9
Whiting	22	Surface @ C/9:	Gloss	Gloss
Barium carbonate	10	Fluidity:	Some	Some
Zircopax	8	Stains show:	Yes	Darks
Zinc oxide	5	Opacity:	Transparent	Translucent
Flint	5	Color:	Clear	Whitish
	100	Note:	Cracks	Cracks

SSG 6 Crackle Clear Glaze

			Reduction	Salted
Kingman feldspar	50	Temperature:	C/9	C/9
Flint	15	Surface @ C/9:	Gloss	Gloss
Whiting	15	Fluidity:	Some	Some
Zinc oxide	15	Stains show:	Yes	Yes
China clay	5	Opacity:	Transparent	Transparent
	100	Color:	Clear	Clear
		Note:	Cracks	Cracks

			Reduction	Salted

SSG 7 Grayish Glaze

Cornwall stone	49	Temperature:	C/9–10	C/9–10
Whiting	25	Surface @ C/9:	Matt	Matt
China clay	23	Fluidity:	Little	Little
Red iron oxide	3	Stains show:	No	No
	100	Opacity:	Opaque	Opaque
		Color:	Broken gray and blues	Broken chocolate and grays
		Note:	Cracks	Cracks

SSG 8 Shaner Yellow-Red (*Studio Potter,* Summer 1974)

Potash feldspar	47.1	Temperature:	C/7–9	C/7–9
Zircopax	22.1	Surface @ C/9:	Semi-matt	Semi-matt
Whiting	19.2	Fluidity:	Some	Some
Talc	3.9	Stains show:	No	No
Bone ash	3.9	Opacity:	Opaque	Opaque
Iron oxide	3.8	Color:	Chocolate brown	Yellow-cream
	100.0			

SSG 9 Greenish Glaze (*Studio Potter*. Summer 1974)

Oxford feldspar	46.4	Temperature:	C/9–10	C/9–10
(Custer OK)		Surface @ C/9:	Gloss	Matt
Edgar plastic kaolin	20.9	Fluidity:	Some	Some
Whiting	11.4	Stains show:	Darks	No
Dolomite	9.0	Opacity:	Translucent	Opaque
Bone ash	4.7	Color:	Greenish	Cream
Tin oxide	3.8			
Talc	1.9			
Iron oxide	1.9			
	100.0			

SSG 10 Blackish Glaze

Soda feldspar	40	Temperature:	C/8–10	C/8–10
Flint	20	Surface @ C/9:	Gloss	Matt and gloss
China clay	16	Fluidity:	Some	Some
Whiting	16	Stains show	No	Darks
Zinc oxide	6	Opacity:	Opaque	Translucent
Red iron oxide	2	Color:	Blackish	Greenish
	100			

SSG 11 Transparent Glaze

Borax, calcined	33.3	Temperature:	C/8–10	C/8–10
China clay	25.0	Surface @ C/9:	Gloss	Gloss
Whiting	16.7	Fluidity:	Some	Some
Soda feldspar	16.7	Stains show:	Yes	Yes
Ball clay	8.3	Opacity:	Transparent	Transparent
	100.0	Color:	Clear	Clear
		Note:	Some cracks	Cracks

SSG 12 Opaque Glaze

Kingman feldspar	30	Temperature:	C/8–10	C/8–10
China clay	22	Surface @ C/9:	Gloss	Matt and gloss
Flint	20	Fluidity:	Little	Some
Zinc oxide	15	Stains show:	Darks	Darks
Dolomite	13	Opacity:	Opaque	Opaque
	100	Color:	Whitish	Whitish and clear

"Easter Egg Jar," Julie Larson;
stoneware, opal luster over blue glaze, gold luster;
12-1/2" H × 10" W.

five

luster

Luster is the thin metallic film covering a glaze. Its use is believed to have originated in eighth-century Egypt, where it was applied on glass. The earliest known method employed a mixture of metals, iron-bearing clay, charcoal, and glue, placed on opaque white glazeware and fired to a low temperature. Several different colored lusters were produced and applied to wall tile, bowls, dishes, and serving utensils. Although the technique spread from Egypt and Mesopotamia to Iran and Afghanistan during the ninth to eleventh centuries, the later quality was inferior. The extent of the use of luster from the eleventh to the late fifteenth century is not known. Majolica ware, made at Gubbio, Italy, from 1520 to 1560, became noted for its gold and ruby metallic colors produced by a highly prized secret process. Little attention was paid to luster from the late sixteenth century to the eighteenth century. Colored lusters were used during the early stage of manufacturing of Bristol Delft ware at Brislington, England. Richard Frank, a potter working at Brislington, is believed to have produced the first English copper luster around 1740. Later, luster was used to decorate inexpensive pottery. As techniques improved, ceramists decorated their work with lusters of pure gold, silver, and other metals. The art of luster became so refined that it was difficult to distinguish silver-lustered tea sets from sterling silver tea sets without handling the pieces.

Lusters are used today in industry to decorate entire pots or the bases of pots and to add flower, foliage, and design motifs. All-over lusters were used until a cheap process of electroplating metals to ceramics was invented.

Types of Luster

Most lusters are produced by depositing a thin metallic film either on the surface (on-glaze luster) or within the glaze (hard luster). *On-glaze* or resinate luster is not a glaze; rather, it is a compound of metallic salt, resin, oil, and turpentine applied to a fired glaze and fired at a low temperature. During firing, the carbon from the resin absorbs the oxygen in the metallic salt to create a more pure metal that bonds to the glaze. Another on-glaze (Arabian) luster is a mixture of metallic salt, ochre, and binder. The mixture is applied to the glaze and fired to 1350° F.; then the kiln is turned off, reduced to 1100°, and cooled to room temperature. After cooling, the surface of the ware is scrubbed and burnished to bring out the luster.

Hard luster, a glaze with metallic salt, is fired like a glaze; however, the firing cycle is different. The kiln is fired to temperature, turned off, reduced to 1000° F., then cooled to room temperature. Some hard lusters do not need reduction, and various temperatures are possible. Hard luster glazes can contain lead, but glazes high in lead will go black if heavily reduced. The metal oxides and salts used for lusters are copper, chrome, silver, uranium, and manganese. During reduction the carbon absorbs the oxygen, converting the metal from a salt or carbonate to a more pure metal and forming a metal deposit on or in the glaze.

Lusters can be metallic, metalized, colored, or colorless (mother-of-pearl). Variations are achieved by applying the luster over light-colored glazes, altering the thickness of application, and applying several different lusters. Experimentation will indicate which lusters to use and will help determine their proper thickness. If the coat is too thick the luster will flake off; if it is too thin no luster effect will be created.

The most common method of lustering is to use a commercially prepared luster known as *Brianchone luster*. Because the reducing agent is incorporated in the luster, a reducing atmosphere is not needed in the kiln. The

"Teapot and Plate," Ron Carlson;
porcelain, clear celadon, engobes, luster, photo decal;
teapot 7-1/2" H.

luster is applied to a glazed form and fired to recommended temperature in an oxidation atmosphere. The composition of the luster is metal, resin, and an organic solvent such as lavender oil. During firing the resin and oil burn out, producing a reduction shield and bonding a thin metal film to the surface. Although Brianchone is the easiest luster to apply, it is less durable than hard lusters.

Hard Luster

Sometimes known as Persian luster, hard luster is an integral part of the glaze, rather than just a deposit on the surface. The glaze surface is generally gloss or semi-gloss, occasionally matt or semi-matt. (The matt surface usually obscures the luster.) The glazes are usually fired from C/012 to C/02, although there is no reason why higher-temperature glazes should not work as well. Many different types of glazes produce lusters, but the lead and alkaline types seem to predominate.

Most formulas call for the clay to be bisque-fired or glazed at a higher temperature

96

than the maturing point of the luster. The clay is vitreous and the glaze is not fluid at the luster firing temperature. The color under the luster is important. White or light-colored clay and glazes show off light-colored lusters most effectively. Some light-colored lusters are transparent; you can achieve unusual effects by using a colored luster over a different-colored glaze, or even a luster over a glaze of the same color. Painting thick-and-thin coats of luster over a gloss glaze produces a ripple color effect.

The clay bodies for hard lusters are porcelain, stoneware, and earthenware that is bisque-fired to C/02 or higher. Some lusters can be used directly on the unglazed clay, and slips or engobes can be used to decorate the clay surface.

The metals commonly used for lusters are bismuth subnitrate, silver nitrate, cobalt carbonate, copper nitrate, and copper carbonate, singly or in various combinations. Reduction converts the metal in the glaze into a more pure state, making the luster an integral part of the glaze. Low-fired lead-based lusters are not recommended for objects that may come in contact with food. Lusters are classified as iridescent (pearlescent, mother-of-pearl), metallic (gun metal), metalized (electro-plated metal effect), and variable (opalescent, rainbow, flash-of-color).

"Lidded Container," Ron Carlson;
porcelain, clear celadon, engobes, luster, photo decal;
6-3/4" H.

"Maki's Boots," Marilyn Levine;
stoneware, leather laces, engobes, luster; 6-3/4" H.

OXIDATION

In the oxidation-fired hard luster technique the glaze is mixed, applied to the bisqued or glazed surface, fired to the recommended temperature, and cooled to room temperature. The only difference between oxidation-fired hard luster and regular glazes is that the luster glaze includes the luster compound. Many oxidation-fired hard lusters are dark brown or black with a metallic finish.

We tested luster formulas on porcelain clay and glaze test pieces. The glaze was weighed, mixed, and painted on the porcelain test piece (bisqueware or glazeware, depending upon the formula). The piece was placed in a test kiln, fired to recommended temperature, and cooled to room temperature. The results, as analyzed and recorded for each formula, are presented in figure 5.1. The formulas are grouped in iridescent and metallic categories.

Figure 5.1

Hard Luster Formulas, Oxidation-fired

Iridescent

L 1 Opalescent Turquoise

Colemanite	34.8	Temperature :	C/08–06
Flint	16.6	Surface	
Plastic vetrox	14.9	@ C/08 :	Gloss
Borax	13.6	Fluidity :	Little
Zinc oxide	9.6	Stains show :	Yes
Soda feldspar	4.8	Opacity :	Transparent
Barium		Color :	Slight
carbonate	2.0		opalescent,
Pearl ash	1.7		pale
Copper			turquoise
carbonate	1.4	Note :	Cracks ;
Whiting	.6		apply a
	100.0		medium
			thickness

"Circles in Square," Maurice Grossman ; stoneware, natural, glaze, luster ; 16'' H.

REDUCTION

The hard luster obtained by reduction has a firing, cooling, and reduction cycle different from that of any other luster or glaze. Control of the firing is highly important. The glaze is mixed, applied to previously glazed ware

Metallic

L 2 Metallic Black
(Long Beach State College, Long Beach, Ca.)

White lead	57.9	Temperature :	C/05
Flint	16.5	Surface	
Red clay	8.3	@ C/05 :	Gloss
Tin oxide	5.8	Fluidity :	Fluid
Copper oxide	4.1	Stains show :	None
Soda feldspar	3.3	Opacity :	Opaque
Cobalt		Color :	Metallic black
carbonate	2.5		
Nickel oxide	.8		
Antimony	.8		
	100.0		

L 3 Metallic Gloss Black

White lead	54.1	Temperature :	Fire until
Frit #5301			gloss
(Ferro)	18.0	Surface :	Gloss
Flint	18.0	Fluidity :	Some
Copper		Stains show :	No
carbonate	3.6	Opacity :	Opaque
Red iron		Color :	Metallic black
oxide	3.6		
Cobalt			
carbonate	2.7		
	100.0		

L 4 Metallic Lead Black
(Long Beach State College, Long Beach, Ca.)

White lead	46.5	Temperature :	C/05
Cornwall		Surface	
stone	35.5	@ C/05 :	Semi-gloss
Flint	5.1	Fluidity :	Some
Albany slip	4.6	Stains show :	No
Manganese		Opacity :	Opaque
dioxide	3.6	Color	
Copper oxide	3.2	/oxidation :	Metallic black
Red iron		/reduction :	Black
oxide	.8		
Burnt umber	.7		
	100.0		

(there are some exceptions, using bisqueware), and fired to the recommended temperature. The kiln is turned off and permitted to cool to about 1200° to 1300° F., depending upon the glaze. At that temperature it is heavily reduced, and the reduction atmosphere is maintained until it cools to 1000°. Then it cools to room temperature at its own rate. You can reduce some glazeware by removing it from the kiln at 1300° and placing it in a lidded container filled with organic matter, a process similar to Raku except that the ware is not placed in water to cool. The ware, covered with carbon, is easily cleaned.

The formulas listed in figure 5.2 were tested on porcelain clay and glaze test pieces. The luster was weighed, mixed, and painted on two porcelain test pieces (bisqueware or glazeware, depending upon the formula). One piece was placed in a test kiln, fired to recommended temperature, cooled accordingly, reduced, and cooled to room temperature. The second test piece was fired in the same way but was not reduced. The results were analyzed and recorded for each formula. The formulas are grouped as iridescent, metallic, metalized, and variable.

Figure 5.2

Hard Luster Formulas, Reduction-fired

Iridescent

L 5 Yellow-Gold Iridescent

Frit #3191		Temperature:	C/06–05
(Ferro)	93	Surface	
Bismuth		@ C/06:	Gloss
subnitrate	5	Fluidity:	Some
Boric acid	2	Stains show:	Yes
	100	Opacity:	Translucent
		Color	
		/oxidation:	Yellow-gold iridescent

Technique: Apply an even coat on glazeware fired to C/5 or higher; fire to C/05; cool to C/018; reduce heavily to 1000° F.; and cool at its own rate.

L 6 Silver Sheen

Frit G 23		Temperature:	C/07–05
(Ferro)	87	Surface	
Tin oxide	8	@ C/06:	Gloss
Silver nitrate	4	Fluidity:	Some
Bismuth		Stains show:	Darks
subnitrate	1	Opacity:	Translucent
	100	Color	
		/reduction:	Silver sheen, iridescent

Technique: Apply to glazeware that has been fired to C/2 or higher; fire to C/06; cool to C/018; reduce heavily and hold reduction to 1000° F.; and cool at its own rate.

L 7 Iridescent Brown

Colemanite	78.2	Temperature:	C/08
Nepheline		Surface	
syenite	15.2	@ C/08:	Gloss
Red iron		Fluidity:	Some
oxide	4.0	Stains show:	Yes
Bismuth		Opacity:	Transparent
subnitrate	1.8	Color	
Resin	.8	/oxidation:	Light brown
	100.0	/reduction:	Brown with a high gloss sheen and slight mother-of-pearl

Technique: Mix, then grind and apply the mixture to a glazed surface. Be sure the application is even. Reduce using Raku technique.

L 8 Opalescent Slight Green

Flint	35.24	Temperature:	C/08–06
Borax	27.36	Surface	
Zinc oxide	19.38	@ C/06:	Gloss
Soda feldspar	9.55	Fluidity:	Little
Barium		Stains show:	Darks
carbonate	3.95	Opacity:	Opaque
Pearl ash	3.32	Color	
Whiting	1.14	/reduction:	Opalescent, green-white
Copper			
carbonate	.06		
	100.00		

Technique: Fire to C/06; cool to C/010; and reduce. Hold reduction to 1000° F.; and cool to room temperature.

Metallic

L 9 Metallic Black

Frit #740 (lead base)	84.4	Temperature:	C/07–04
Kaolin	12.6	Surface	
Cobalt carbonate	3.0	@ C/04:	Matt
	100.0	Fluidity:	None
		Stains show:	None
		Opacity:	Opaque
		Color	
		/oxidation:	Matt black
		/reduction:	Metallic matt black

Technique: Apply even coat; heavy reduction.

L 10 Metallic Blue-Black

Red lead	59.2	Temperature:	C/04
Flint	21.6	Surface	
China clay	11.8	@ C/04:	High gloss
Cobalt oxide	2.1	Fluidity:	Some
Soda feldspar	1.8	Stains show:	Darks
Manganese dioxide	1.3	Opacity:	Translucent = thin; opaque = thick
Copper oxide	1.3		
Black iron oxide	.9	Color	
	100.0	/oxidation:	Blue-black
		/reduction:	Metallic blue-black

Technique: Fire to C/04; reduce; hold reduction to C/09; cool to room temperature.

L 11 Metallic Browns

Borax	56	Temperature:	C/08
Flint	26	Surface	
Colemanite	11	@ C/08:	Gloss
Manganese carbonate	5	Fluidity:	Little
Red iron oxide	2	Stains show:	Yes
	100	Opacity:	Translucent = thin; opaque = thick
		Color	
		/oxidation:	True brown
		/reduction:	Metallic brown

Technique: Fire to C/08; turn off kiln; reduce for half an hour.

Metalized

L 12 Metalized Silver and Golds

Colemanite	78.0	Temperature:	C/04
Cornwall stone	19.0	Surface	
Silver nitrate	1.8	@ C/04:	Gloss
Tin oxide	.8	Fluidity:	Some
Uranium, yellow	.4	Stains show:	None
	100.0	Opacity:	Opaque
		Color	
		/oxidation:	Slight yellow
		/reduction:	Silver and golds, lustrous

Technique: Grind mixture; reduce heavily as in Raku.

L 13 Metalized Olive Luster

Colemanite	72.4	Temperature:	C/06
Cornwall stone	18.0	Surface	
Ochre, yellow	7.2	@ C/06:	Gloss
Black copper oxide	1.8	Fluidity:	None
Cobalt carbonate	.6	Stains show:	Darks
	100.0	Opacity:	Translucent
		Color	
		/oxidation:	Transparent green
		/reduction:	Olive green, luster

Technique: Grind mixture; heavy reduction as in Raku.

L 14 Metalized Old Silver

Colemanite	60.7	Temperature:	C/06
Calcined borax	23.4	Surface	
Nepheline syenite	4.7	@ C/06:	Gloss
Kaolin	4.7	Fluidity:	Little
Rutile	4.7	Stains show:	None
Black copper oxide	1.8	Opacity:	Opaque
	100.0	Color	
		/reduction:	Old silver and blues, lustrous

Technique: Heavy reduction as in Raku.

"Cups," Ron Carlson; porcelain, engobes, celadon glaze, lusters, photo decals; 4'' to 6'' H.

L 15 Metalized Brown

Colemanite	60.0	Temperature:	C/010
Nepheline syenite	25.5	Surface @ C/010:	Gloss
Lithium carbonate	10.0	Fluidity:	Little
Copper carbonate	3.0	Stains show:	No
Manganese dioxide	1.5	Opacity:	Opaque
	100.0	Color	
		/oxidation:	Black
		/reduction:	Rich, lustrous brown

Technique: Heavy reduction as in Raku; best over previously glazed gloss surfaces.

L 16 Metalized Browns

Red lead	60.0	Temperature:	C/010
Copper oxide	9.0	Surface	
Soda feldspar	7.6	@ C/010:	Gloss
Flint	7.6	Fluidity:	Little
Manganese dioxide	6.0	Stains show:	No
Cobalt oxide	3.5	Opacity:	Opaque
China clay	2.9	Color	
Whiting	2.9	/oxidation:	Metallic black
Red iron oxide	.5	/reduction:	Lustrous browns
	100.0		

Technique: Turn off kiln; reduce for half an hour; best over previously glazed gloss surfaces.

L 17 Brilliant Red

White lead	50	Temperature:	C/06
Potash feldspar	24	Surface @ C/06:	Gloss
Flint	14	Fluidity:	Some
Cadmium selenium stain	5	Stains show:	Yes
		Opacity:	Translucent
Kaolin	4	Color	
Copper nitrate	3	/reduction:	Slight metalized, brilliant red
	100		

Technique: Apply on bisqueware or glazeware fired to C/2 or higher; fire to C/06; reduce at C/018; hold reduction to 1000° F.; and cool to room temperature.

L 18 Metalized Oil

Frit #14	48.3	Temperature:	C/07
Colemanite	48.3	Surface	
Black copper oxide	2.4	@ C/07:	Gloss
Manganese dioxide	1.0	Fluidity:	Some
	100.0	Stains show:	Darks
		Opacity:	Oxidation = transparent; reduction = opaque
		Color	
		/oxidation:	Transparent steel blue
		/reduction:	Oiled, multi-colored luster

Technique: Heavy reduction as in Raku; best over previously glazed gloss surfaces.

L 19 Metalized Golds

Flint	36.1	Temperature:	C/05
Cryolite	11.9	Surface	
Soda feldspar	11.9	@ C/05:	Semi-gloss
Tin oxide	9.6	Fluidity:	Little
Lithium carbonate	8.1	Stains show:	No
Bone ash	6.9	Opacity:	Opaque
Whiting	5.3	Color	
Bismuth subnitrate	3.9	/oxidation:	Light blue
China clay	3.5	/reduction:	Lustrous old golds
Bentonite	1.5		
Silver nitrate	1.3		
	100.0		

Technique: Heavy reduction as in Raku; best over previously glazed gloss surfaces; apply even coating.

L 20 Metalized Blue Silver

Boric acid	27.5	Temperature:	C/04
Whiting	17.4	Surface	
Flint	16.1	@ C/04:	Gloss
Kaolin, calcined	12.9	Fluidity:	Some
Litharge	11.8	Stains show:	Yes; reduction = darks
Soda ash	10.1		
Bismuth subnitrate	2.8	Opacity	Transparent; reduction = translucent
Copper sulfate	1.1		
Cobalt carbonate	.3	Color	
	100.0	/oxidation:	Transparent light true blue
		/reduction:	Smooth blue-silver

Technique: Heavy reduction as in Raku.

Variable

L 21 Variable Grays and Silvers

Frit #740	70	Temperature:	C/05
Kaolin	10	Surface	
Whiting	8	@ C/05:	Semi-gloss
Tin oxide	5	Fluidity:	Little
Cobalt oxide	4	Stains show:	No
Manganese		Opacity:	Opaque
dioxide	3	Color	
	100	/oxidation:	Black
		/reduction:	Lustrous dark blue, gray, and silver

Technique: Apply even coat on glazed surfaces; reduce heavily.

L 22 Variable Yellows and Whites

Colemanite	68	Temperature:	C/05
Plastic vetrox	25	Surface	Gloss to
Tin oxide	5	@ C/05:	semi-gloss
Silver nitrate		Fluidity:	Little
(dry)	2	Stains show:	Darks
	100	Opacity:	Translucent
		Color	
		/oxidation:	Grayed tan
		/reduction:	Pearly yellows and whites

Technique: Fire until gloss and smooth; reduce heavily.

L 23 Variable Yellowish Brown

White lead	67.2	Temperature:	C/04
Albany slip	28.0	Surface	
Copper		@ C/04:	Semi-gloss
carbonate	2.8	Fluidity:	None
Bismuth		Stains show:	Darks
subnitrate	2.0	Opacity:	Translucent
	100.0	Color	
		/oxidation:	Dark amber
		/reduction:	Yellowish brown with slight iridescence

Technique: Reduce heavily as in Raku.

"Lidded Jar," Charles Counts; stoneware, glazes and luster; 11-1/2" H × 7-1/2" W.

L 24 Variable Base Glaze

Colemanite	60	Temperature:	C/010
Nepheline		Surface	
syenite	27	@ C/010:	Gloss
Flint	8	Fluidity:	Little
Tin oxide	5	Stains show:	Yes
	100	Opacity:	Translucent
		Color	
		/oxidation:	Whitish
		/reduction:	Whitish

Copper sulfate	4	Copper flash, slight blue
Silver nitrate, dry	4	Old golds and silver flash
Chromium nitrate	2	Olive greens, slight luster
Bismuth subnitrate	4	Gray, silver flash
Uranium, yellow	4	Grayed yellow, slight luster
Potassium chromate	4	Medium grayed yellow-green, slight luster
Copper phosphate	4	Turquoise, copper, and blue flash

Technique: Fire to C/010 and reduce heavily to 1000° F.

L 25 Variable Copper Green and Red

Red lead	50.0	Temperature:	C/04
Flint	30.8	Surface	
Boric acid	7.5	@ C/04:	Semi-gloss
Pearl ash	5.6	Fluidity:	None
Copper		Stains show:	Darks
carbonate	2.8	Opacity:	Opaque
Silver nitrate,		Color	
dry	1.5	/oxidation:	Green-black
Zinc oxide	.9	/reduction:	Flashes of
Tin oxide	.9		green, red,
	100.0		copper

Technique: Reduce heavily; close damper; cool slowly. Grind; apply as a glaze; use a small amount of glue.

L 26 Variable Brown Bronze

Litharge	48.6	Temperature:	C/06
Silica	32.2	Surface	
Kaolin,		@ C/06:	Gloss
calcined	11.3	Fluidity:	None
Soda ash	3.7	Stains show:	Darks
Bismuth		Opacity:	Transparent;
subnitrate	2.7		reduction =
Copper			translucent
sulfate	1.5	Color	
	100.0	/oxidation:	Light sea-
			foam green
		/reduction:	Broken blues
			and old
			brown
			bronze

Technique: Apply even coat; grind glaze; reduce heavily.

L 27 Variable Greens

Red lead	47.9	Temperature:	C/08
Cornwall		Surface	
stone	19.0	@ C/012:	Gloss
Cullet	18.0	Fluidity:	Little
Flint	9.6	Stains show:	Yes
China clay	2.4	Opacity:	Transparent
Silver nitrate,		Color	
dry	1.2	/oxidation:	Yellow-green
Bismuth		/reduction:	Greens and
subnitrate	1.1		silvers
Copper			
carbonate	.8		
	100.0		

Technique: Grind mixture; reduce heavily as in Raku.

L 28 Variable Brown

Lead		Temperature:	C/04
bisilicate	41.2	Surface	Semi-gloss
Buckingham		@ C/04:	to gloss
feldspar	12.4	Fluidity:	Little
Flint	12.0	Stains show:	No
Copper		Opacity:	Opaque
carbonate	11.2	Color	
Tin oxide	9.4	/oxidation:	Medium
Kaolin	6.9		green
Whiting	5.6	/reduction:	Lustrous
Titanium	1.3		brown
	100.0		

Technique: Grind mixture; apply even coat; best over a glazed gloss surface.

L 29 Variable Old Silver

Litharge	40.9	Temperature:	C/06
Silica	24.1	Surface	
Whiting	17.1	@ C/06:	Semi-matt
Kaolin,		Fluidity:	None
calcined	13.4	Stains show:	Yes;
Bismuth			reduction
subnitrate	3.1		= no
Copper		Opacity:	Transparent;
sulfate	1.4		reduction
	100.0		= opaque
		Color	
		/oxidation:	Mint green
		/reduction:	Broken old
			silver, blues,
			and greens

Technique: Grind; apply even coating on surface; best over glazed gloss surfaces.

L 30 Variable Light Blue and Green

Boric acid	32.1	Temperature:	C/04–1
Whiting	20.3	Surface	
Flint	20.1	@ C/04:	Semi-gloss
Kaolin,		Fluidity:	Some
calcined	16.5	Stains show:	Yes;
Soda ash	8.8		reduction
Copper			= darks
sulfate	1.9	Opacity:	Transparent;
Cobalt			reduction =
carbonate	.3		translucent
	100.0	Color	
		/oxidation:	Light true
			blue
		/reduction:	Broken light
			blue and
			greens, slight
			mother-of-
			pearl

Technique: Fire to C/04; cool to C/010; reduce to C/017; let cool to room temperature.

L 31 Variable Silver, Blue, and Green

Flint	32.0	Temperature:	C/05
Red lead	31.1	Surface	
Borax	19.2	@ C/05:	Semi-gloss
Soda feldspar	10.5	Fluidity:	None
Kaolin	2.0	Stains show:	None
Boric acid	2.0	Opacity:	Opaque
Barium		Color	
chloride	1.8	/oxidation:	Broken light
Copper sulfate	.5		jade
Bismuth		/reduction:	Silver, green,
subnitrate	.5		and blue
Silver nitrate	.2		lusters
Silicon			
carbonate	.2		
	100.0		

Technique: Apply even coating; reduce heavily as in Raku.

L 32 Variable Blues and Golds

Red lead	32.0	Temperature:	C/09
Soda feldspar	19.5	Surface	
Flint	19.0	@ C/09:	Semi-gloss
Borax	19.0	Fluidity:	Some
Bismuth		Stains show:	Yes
subnitrate	4.5	Opacity:	Transparent
Pearl ash	2.0	Color	
Kaolin	2.0	/oxidation:	Medium
Potassium			yellow
dichromate	2.0	/reduction:	Medium
	100.0		yellow with
			lustrous
			blues and
			golds on
			surface
		Note:	Cracks

Technique: Reduce heavily as in Raku.

L 33 Variable Rose and Silver

Pearl ash	29.4	Temperature:	C/04–02
Flint	22.9	Surface	
Kaolin,		@ C/04:	Semi-gloss
calcined	18.8	Fluidity:	None
Boric acid	12.2	Stains show:	Yes
Whiting	5.7	Opacity:	Transparent;
Bismuth			reduction =
subnitrate	3.8		translucent
Soda ash	2.9	Color	
Zinc oxide	2.9	/oxidation:	Slight blue
Copper		/reduction:	Broken rose
sulfate	.9		and silver
Silver nitrate,			
dry	.5		
	100.0		

Technique: Grind; apply even coating; reduce heavily as in Raku.

L 34 Variable Silvers

Feldspar		Temperature:	C/04
(soda)	24.2	Surface	
Borax	23.3	@ C/04:	Gloss
White lead	17.5	Fluidity:	Little
Flint	11.6	Stains show:	Darks
Whiting	10.5	Opacity:	Opaque
Tin oxide	8.4	Color	
Silver nitrate,		/reduction:	Variable
dry	3.2		silvers
Soda ash	.7		
Kaolin	.5		
	99.9		

Techniqe: Fire over previously fired glaze; cool to C/010; reduce heavily as in Raku.

Fuming

Fuming is a method of producing lusters by placing a metallic compound in the kiln, where it vaporizes. The vapors coat everything in the kiln with a thin luster film. Depending upon the glaze, temperature, clay body, and compound used, the resulting film will give a pearlescent, mother-of-pearl, blush-of-color, or iridescent effect. Some effects will be so subtle that the pot must be examined in bright sunlight to detect the luster; others will have obvious, distinctive blushes of color.

For fuming, use the following chemicals by themselves or in combination: iron chloride, copper sulfide, silver nitrate, mercury sulfide, bismuth subnitrate, zinc nitrate, sodium nitrate, strontium nitrate, barium nitrate, chrome nitrate, cupric chloride, and stannous chloride. Stannous chloride by itself gives the most effective pearlescent luster; 60 to 80 percent stannous chloride with various combinations of the other chemicals also gives good results.

PROCEDURE

Most clay bodies and glazes, from earthenware to porcelain, can be fumed. Ware for fuming is made, glazed, and handled the

"Lidded Box," Jean Yates;
stoneware, metal handle, luster, oxides; 10-1/2" H.

same as ware for any other process. The ware is loaded into the kiln and glaze-fired to temperature, using either oxidation or reduction. The kiln is cooled at its own rate to 1400° F., and at this temperature the dampers, peepholes, and openings in the kiln are closed so that the temperature inside cools to a uniform 1300°. Uneven temperatures give uneven results. (A pyrometer is desirable but not necessary.) At 1300° the objects in the kiln are barely discernible.

The pilot lights, burners, or blowers are turned on at a low setting to help spread the fumes. The air movement helps produce an overall effect, rather than flashings of iridescence. With a long spoon the compound is inserted into the porthole, fire box, peep-hole, or if necessary down the stack. The purpose is to spread as much of the compound as possible around the kiln. Several small spoonfuls of the compound will produce better results than one large one. The compound will burst into fumes that the blowers will spread around, soaking the atmosphere. Remove draw rings

to determine the amount of iridescence and whether additional luster is necessary. If the compound falls directly on the surface of the glaze, a white or gray film may result that is difficult to remove. After the fuming is completed, the burners, pilot lights, and blowers are turned off and the kiln cools at its own rate.

PAN FUMING

The major disadvantage of fuming is that everything in the kiln will take on iridescence. To fume individual pieces glassblowers use pan fuming, a technique suitable for the ceramist. On a hotplate, heat a metal pan containing the fuming compound. When the kiln has cooled to dull red heat, remove a ceramic piece from the kiln with tongs. Hold the form over the fuming chemicals, rotate it several times, and return it to the kiln to cool to room temperature at the kiln's rate. Since the temperature of the piece removed from the kiln is above the quartz inversion temperature, breakage is rare. However, some clay bodies will not take the thermal shock of pan fuming.

SAFETY

Because metal fumes are toxic to both the eyes and the lungs, a kiln with a hood or chimney is desirable. If you use an electric kiln without a hood indoors, you may need a gas mask because the heavy fumes will seep from the kiln or the pan into the room. Because the compound may cause a skin rash, handle it with a spoon instead of with your fingers.

On-Glaze Luster

On-glaze luster is applied to the surface of a glaze. The metals are set on the glaze, the piece is fired, and the metal bonds to the surface. *Resinate* luster uses a compound containing a reducing agent, and the firing is done by oxidation. The *Arabian* compound does not contain a reducing agent, because the reduction is done in the kiln or in a metal container.

ARABIAN LUSTER

Mix metal carbonates or salts with yellow ochre and glue. The ochre acts as a medium, and the glue helps hold the mixture to the glazed surface. The glaze should have a gloss finish and be fired to C/06 or C/07. Grind the mixture, then spray or paint it on a fired, glazed surface. Fire the luster to the temperature at which the previously fired glaze starts to become soft, around 1200° F. Reduce the kiln heavily to the point that smoke comes out the portholes. Then turn down the burners, but maintain a reduction atmosphere until 1000°. Turn the burners off and let the kiln cool at its own rate. After cooling, wash off the mixture; you may need a mild abrasive.

The luster color and iridescence depend upon the glaze, the duration of the reduction, and the luster composition. Lead-based white or light-colored glazes containing tin give the best results. The metals used are bismuth subnitrate, silver carbonate, copper sulfate, silver nitrate, stannous oxide, and mercury sulfide.

A variation of the reduction technique for Arabian luster is to reduce outside the kiln, as in Raku reduction. Apply the luster mixture to a previously glaze-fired form, and fire to the recommended temperature. Turn off the kiln and allow it to cool to 1300° F. Remove the object from the kiln; place it in a metal container partly filled with shredded wood, wood dust, dry leaves, or crumpled newspaper; and close the lid tightly. The organic material will ignite, causing a heavy, carbon-filled atmosphere. Leave the piece in the container until it reaches 500°; then remove it. (If you remove it too soon, the luster may re-oxidize and be lost.) Remove any carbon deposits by polishing with a mild abrasive.

The formulas in figure 5.3 were tested on porcelain clay and glaze test pieces. The luster was weighed out, mixed, and painted on a previously glazed porcelain test piece. The test piece was placed in a test kiln, fired to the recommended temperature, reduced, and cooled. The results were analyzed and recorded for each formula.

Figure 5.3

Arabian Luster Formulas

L 35 Slight Blue Luster

Red ochre	85	Temperature:	C/019
Bismuth		Color	
subnitrate	12	/reduction:	Slight blue,
Silver			variable
carbonate	3		
	100		

Technique: Apply over ware that was glazed with a high-gloss lead glaze and fired to C/07 or higher. Fire luster to C/019; reduce; hold reduction for half an hour. Cool to room temperature.

L 36 Variable Silverish

Yellow ochre	80	Temperature:	1210° F.
Bismuth		Opacity:	Transparent
subnitrate	16	Color	
Silver nitrate,		/reduction:	Silverish
dry	4	Note:	Try on
	100		colored
			glazes

Technique: Grind mixture with water and a small amount of glue. Paint mixture on piece that has been glaze-fired to 1800° F. or higher with a gloss glaze. Fire to 1210° F., then reduce heavily and hold the reduction to 1000° F. After cooling, remove the film with a mild abrasive (scrub brush) to bring out the luster.

L 37 Variable Reds, Yellows, and Greens

Yellow ochre	71.98	Temperature:	1210° F.
Copper		Opacity:	Transparent
sulfate	26.87	Color	
Silver nitrate,		/reduction:	Reds,
dry	1.15		yellows,
	100.00		and greens

Technique: Same as for L 36.

L 38 Reddish Sheen

Red ochre	70	Temperature:	C/017
Copper		Color	
carbonate	30	/reduction:	Slight reddish
	100		sheen

Technique: Apply over high-gloss, lead-based glazeware fired to C/07 or higher. Fire luster to C/019; reduce; and hold reduction for half an hour while cooling.

The following two formulas are a slight variation, in that no ochre is part of the compound. The compound is similar to a stain and is used like the other Arabian lusters.

L 40 Orange-Red

Stannous		Temperature:	C/019
oxide	53.3	Color	
Copper sulfide		/reduction:	Thin =
(sulfate)	46.7		yellows;
	100.0		thick =
			orange-
			reds

Technique: Grind mixture with water and glue; paint mixture on glazeware fired to 1800° F. or higher. Fire to 1205° F., then reduce and hold reduction to 1000° F. After cooling, remove film with mild abrasive.

L 41 Variable Yellow

Mercury		Temperature:	1210° F.
sulfide	48.5	Opacity:	Transparent
Copper		Color	
sulfide	48.5	/oxidation:	Tans, yellows
Silver nitrate,		/reduction:	Yellows to
dry	3.0		dark
	100.0		amber;
			thick =
			goldstone

Technique: Mix and grind the mixture; apply thin wash over a gloss-glazed surface that has been fired to 1800° F. or higher; fire to 1210°. Heavily reduce as in Raku.

L 39 Metallic Matt Brown

Copper		Temperature:	C/04–1
carbonate	45	Surface	
Yellow ochre	43	@ C/03:	Matt
Mercury		Fluidity:	None
sulfide	10	Stains show:	None
Silver nitrate,		Opacity:	Opaque
dry	2	Color	
	100	/oxidation:	Metallic black
		/reduction:	Metallic
			brown

Technique: Apply even coating on glazed surface; reduce heavily.

RESINATE LUSTER

Before the invention of electroplating, the resinate luster technique was the most successful method of producing an all-over thin film of metal on the surface of a glaze. It is still used, but primarily for creating a decorative treatment on a small area.

Metallic salts are heated with resin and oil (for example, lavender oil) and thinned with turpentine. The solution is stirred until mixed, allowed to rest for one day, filtered, and decanted after five to ten days. The solution is sprayed or painted on a glazed surface. The firing is to red heat, about 1200° to 1370° F. A thin film of the dissolved metal is deposited on the surface of the glaze, using an oxidation firing. The resin provides a local reduction (on the glaze) so that reducing the entire kiln is unnecessary. Resinate lusters can be fired with other glazes that mature at the same temperature.

"Slab Sculpture," Gerry Williams;
stoneware, photo resist, contact image over white slip;
19-1/2" H.

six

photosensitized ceramics

Ceramics, like any other art form, has several ideals: to create beauty, to use the principles of form to create a coherent structure, and to communicate feelings and ideas. The virtue of clay lies in its simplicity and its direct, instinctive appeal. It is difficult to reduce the inherent beauty of clay forms to some conceptual formula. While the classic shapes of Greek, Roman, Chinese, and Japanese traditional ceramics offer some models of form and design, contemporary ceramists translate into clay their own ideas and feelings. Thus their art communicates their unique interpretation through particular shapes and surfaces. The individual approach may mean a departure from traditional ceramics—from a glazed functional pot, for example—to something unique in which function may be secondary.

In search of individual identification and new approach, many ceramists have explored the particularly intriguing technique of combining clay and photography. Photography, the image-maker, is used to communicate on the surface of the clay form. The principle of sensitizing a nonpaper surface is old.[1] A gelatin binder, holding the silver emulsion, is applied to the surface of the clay form. During development the sensitized silver is converted by the light into black colloidal silver to form the image. The unexposed silver is dissolved away in the fixing. A protective coating is then applied to the surface.

The process of photosensitized ceramics is the most complex of the ceramic processes, for it requires a knowledge of both ceramics and darkroom photographic techniques. The basic photographic skills needed are the ability to use an enlarger and processing developer, a stop-bath, and a fixer. Aside

---

[1] A fascinating historical note is that Thomas Wedgwood, youngest son of Josiah Wedgwood the potter, was familiar with the camera obscura, which his father used as an aid in drawing designs and scenes for ceramic use. Thomas experimented by making pictures with the camera obscura and using silver for direct printing on paper and leather; he almost became the inventor of photography. Thirty years later, in 1826, the first permanent camera image was recorded.

from the enlarger (or slide projector) and safelight, most of the equipment can be found in the kitchen or ceramics studio. We recommend using potsherds to test for the proper thickness of emulsion, exposure time, and processing.

Definition of Terms

Before we examine the equipment and procedure, it will be helpful to review some of the terms used in photosensitized ceramics. Many of the terms are already familiar, but a refresher will help relate them to the photosensitized technique.

"Sleeping Child Burn," John C. Barsness;
white stoneware, oxides,
clear glaze, photo emulsion;
14-1/4" × 16".

Ceramic pigment. A pigment made of ceramic stain, china paint, or metallic oxide. The pigment is mixed with the developer solution or applied after the fixing solution procedure is complete, depending upon the process used. To make the pigment permanent requires firing the finished image with pigment in a kiln with a temperature range of C/018 to C/2, depending upon the pigment used.

Coating. The thin film of photographic emulsion that is applied to the ceramic surface.

Contact transparency. A printing method in which the negative is placed in contact with the sensitized ceramic surface and held in position with tape. The negative-covered area is then exposed to raw light, and the emulsion exposed by light passing through the negative densities.

Developer. The solution that produces the silver image in the normal photographic process. Developer is a water solution containing the chemicals necessary to develop only the exposed silver halide grains in the emulsion, leaving the unexposed grains unaffected.

Emulsion. The light-sensitive coating that is applied to the ceramic surface.

Emulsion speed. The relative sensitivity of emulsion to light. The type of light, distance of light from emulsion, and time of exposure are stated in the manufacturer's directions.

Enlarger. A device for projecting the image of a negative onto the sensitized surface. (Slide or opaque projectors will suffice.) All enlargers can be adjusted to make projections larger than the negative image.

Exposure. The act of subjecting the photosensitive surface to the action of light for a certain period of time.

Fast enlargement. The time required to expose the sensitized surface. Most exposures are about one minute (fast); some are four or more minutes (slow).

"Container #II," John C. Barsness;
white earthenware, oxides,
clear glaze, photo emulsion;
9-3/4" H.

Fix. To make the exposed and developed surface insensitive to further exposure to light. The solution used in this process is called a *fixing bath.*

Gelatin. A protein substance made from animal hooves, used as a chemical-bearing medium and binder.

Halftone. A term for an image that has been screened to produce the effect of continuous tone. This is done by breaking up the image into halftone dots in a regular pattern too fine to be seen without magnification.

Image. The photographic representation of the subject.

Light source. A controlled source of light to project an image for enlarging or contact printing. The light can be ultraviolet, sun, photoflood, or carbon-arc.

Masking. Covering or blocking out a portion of the image area and clay surface with opaque tape or paint, to prevent undesired "fogging" (image developed on the sensitized surface).

Negative. Any photographic image in which the subject tones have been reversed. The negative may have continuous gray areas or just black and white areas.

Projected transparency. A negative that is projected onto the sensitized ceramic surface using an enlarger, a slide projector, or an opaque projector.

Projector. A machine used to project an enlarged image onto a sensitized surface. (Areas not to be exposed are masked out.) Slide or opaque projectors are the most commonly used.

Positive. An image in which the tones are similar to those of the subject.

Safelight. Special color illumination (usually yellow or red) used in the darkroom that does not appreciably affect photosensitized emulsions.

Sensitized surface. A ceramic surface to which has been applied an emulsion sensitive to light energy.

Stop bath. Acid rinse bath that follows the developing step and stops any further development of the sensitized surface.

Surface. The area to which the emulsion is applied and which will be exposed to light to produce a photographic image.

Transparency. A transparent image viewed by transmitted light. The black areas are opaque and project no image; the clear areas project light; the gray areas project light in direct correlation to the amount of gray in the transparency.

Vitrifaction. Providing ceramic ware with a glassy coating or glaze to make the surface extremely low in porosity. The vitrified surface prevents the photographic emulsion from seeping into the pores of the clay and permits the proper developing and fixing of the image.

Equipment and Materials

1. Fired glazed clay form (if unglazed, the form must have a smooth, nonporous surface).
2. Enlarger or slide projector.
3. Emulsion (described below in detail). Commercial emulsions available are Kodak KPR (Kodak Photo Resist), KOR (Kodak Ortho Resist); Vandyke Sensitizer formula; Kallitype Sensitizer formula; Rockland CB 101, BB 201, BX 201, DTE; Picceramic.
4. Darkroom equipment: measuring cup, storage jar, safelight, sponge, trays, etc.
5. Photographic developing chemicals: developer, stop bath, and fixer.
6. Light-tight working area.
7. Light source for contact prints: ultraviolet light, sun, photoflood.
8. Ceramic pigment: china paint, underglaze, stains.
9. Positive transparency: halftone or solid black.
   A. Contact transparency for Kallitype, Vandyke, Kodak KPR, and Rockland BX 201.
   B. Projected transparency for Kodak KOR; Rockland CB 101, DTE, and BB 201; and Picceramic.
10. Protective finish for silver-based emulsions: varnish, lacquer.

Surface

Almost any flat or slightly curved ceramic surface can be used for this technique, including glazed, hard bisque, glass, enamel, and even salt-glazed. Dry greenware and porous bisqueware can be used with the ceramic print emulsions, providing that before the emulsion is applied the surface is sealed with a vegetable or decoupage glue. Dark-colored surfaces will obscure the finished transparent image; however, because the texture and color of the base surface will still be visible, this can be an effective treatment. In general, light colors and surfaces of gloss, semi-gloss, and semi-matt seem to work best.

The ceramic print emulsions (Kodak KPR, KOR, and Picceramic emulsions) require a ceramic pigment in the developing and fixing process. The pigment remains on the surface of the glass, clay, enamel, or glaze. It is fired to C/016 or higher to become a vitrified, integral part of the surface.

Emulsions

Kodak, Rockland, Picceramic, and other manufacturers prepare emulsion kits and mixes. Kodak's booklet AJ-5 lists procedures and formulas for making Vandyke and Kallitype emulsions. It is difficult to predict how good the image will be, because the results may be affected by various factors: the surface of the glaze, clay, enamel, or glass; variation in the absorption rate and hardness; problems of using the enlarger to project the image onto curved surfaces; and development of the image on large and awkwardly shaped pots.

Listed below are various commercial emulsions. All the silver- and silver nitrate–based emulsions require an additional protective finish to prevent the finished image from being scratched.

Vandyke Sensitizer is a silver nitrate–based emulsion that is custom-formulated by the individual user. The chemicals are available from local chemical houses and are not difficult to prepare. Once prepared, the emulsion is stored in a brown glass bottle out of direct light, where it will keep for several months. The formula and additional information are found in the Kodak booklet AJ-5.

Kallitype Sensitizer is a silver nitrate–based emulsion, more complex than the Vandyke but also custom-formulated by the individual ceramist. The quality of the finished photograph is better with Kallitype than with Vandyke. The oxalic acid and silver nitrate in the formula may cause skin irritation and burns, so use caution in handling these and other photographic materials. For the formula for Kallitype Sensitizer and additional information, see Kodak booklet AJ-5.

Rockland CB 101 is a medium-fast, prepared emulsion for all-around use with black tones that can be used with enlargement and contact printing. A silver-based emulsion, it is recommended for enlargements.

Rockland BB 201 is a fast enlargement, prepared, silver-based emulsion.

Rockland BX 201 is a low contact-speed, prepared, silver-based emulsion with a high-contrast image that requires a bright light for exposure.

Picceramic is an enlargement emulsion available in kit form that can be painted, poured, or sprayed onto the ceramic surface. The kit contains the *materials* necessary to make up the emulsion for a ceramic print image. The image is permanently adhered to the clay surface when fired in the kiln.

Kodak KPR is a contact, prepared emulsion that uses ceramic pigment to permanently adhere the image to the ceramic surface. Various ceramic pigments can be used, such as stains, china paints, or overglazes. N-butyl acetate is the developer, and adequate ventilation is advised.

Kodak KOR is a prepared emulsion, recommended for enlargement, that uses ceramic pigments for a vitrified surface. It is more difficult than KPR to use and can be less stable. A slide projector or enlarger can be used to project the positive transparency onto the prepared ceramic surface.

Rockland DTE is a prepared enlargement-speed, silver-based emulsion that has been dehydrated for indefinite storage.

Emulsion Application

Just as photographers use test strips in developing photographic prints, ceramists using the photosensitizing process test on potsherds. Testing determines the proper thickness of emulsion, exposure time, development process, and finishing.

Some emulsions are available premixed in kit form, but the Vandyke and Kallitype Sensitizers require the purchase of the individual ingredients to make up the emulsion. Regardless of the emulsion system you use, the total cost is reasonable considering the result. Rockland emulsions are premixed and have a long shelf life, while the pigment and stabilizer mix for the Picceramic process lose strength in less than three months. Pay careful attention to the directions that accompany the emulsions, because there is little latitude allowed.

Clean the surface of the ceramic form of all dust and dirt. You can then brush, pour, or spray the emulsion onto the dry surface. Rockland DTE is especially formulated for spray application, but the spray is irritating and the darkroom must be well ventilated; use a mask.

The emulsion is applied in a red safelight area and should not be exposed to external light. Two thin coats give the best results; with high-contrast negatives, however, one coat may be sufficient. Spread the solution horizontally; then rework with vertical strokes. Remove excess sensitizer and allow the form to dry in a dark, dust-free place.

If the emulsion and/or the ceramic surface are cold, the emulsion may not adhere properly. Place the bottle of emulsion in a bowl of hot water until the emulsion temperature reaches 75° F. Warm the clay form in an oven or kiln until the temperature is 90 to 100°. Do not permit the temperature of the emulsion or the clay to exceed 105°.

Silver nitrate and oxalic acid stain hands

and clothing. Avoid contact with the emulsion by wearing plastic or rubber gloves and an apron.

Exposure

The transparency used to expose the design onto the emulsion is generally a positive, as opposed to a negative. (See the section on positives in chapter 2 for more details.) The enlargement or contact positive must have clean, clear black-and-white images—either halftone or solid blacks. Thin transparencies (translucent blacks) do not print well, even though some of the emulsions have a high contrast.

Contact transparency is a transparency the same size as the planned, printed area that comes in contact with the clay surface. The application of emulsion, storing of sensitized ceramic forms, and exposure and development are all done under darkroom conditions using a red safelight. The contact transparency is positioned on the sensitized surface for close contact. With flat surfaces such as tile, you will need only a sheet of glass to hold the transparency in place. Slightly curved or irregular surfaces are more difficult to work with and require the use of tape to hold the transparency. The light is then positioned over the transparency (fig. 6.1).

The exposure time varies depending on the brand and batch of emulsion and the light source. In general, exposure time ranges from two to ten minutes for sunlight, photoflood, carbon-arc, or ultraviolet light. (When you use ultraviolet light, be sure to wear protective eyeglasses.)

Figure 6.1

Exposing sensitized clay surfaces with projected transparency (left)
and contact transparency (right): (A) enlarger or slide projector;
(B) transparency; (C) sensitized surface;
(D) clay form; (E) light source; (F) glass plate.

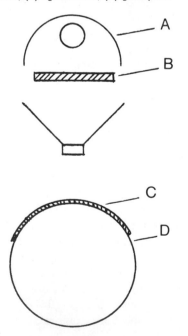

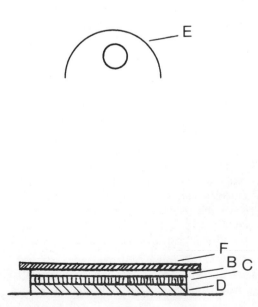

Enlargement transparency is a positive to be projected on the sensitized ceramic surface. The quality of the transparency is important. Since the enlargement transparency is smaller than the contact transparency, the translucent blacks not visible without magnification will show up under the enlarger. Halftone or solid black transparencies can be used in various types of enlargers or slide projectors that hold 35mm, $2\frac{1}{4}''$, or larger-format transparencies. Areas around the image are masked so that only the image will "take." Several different projections can be made on the surface by masking off the areas to create multi-images. Both flat and slightly curved surfaces can be used (see figure 6.1). All projections must be done under darkroom conditions, using a red safelight.

The exposure time ranges from 30 seconds (Picceramic) to eight minutes (Kallitype emulsion). Testing on rejects is recommended, because the exposure time varies with the brand, batch of emulsion, and light source employed.

Development

The manufacturer recommends a specific developer, fixer, temperature, and time of development for each emulsion. Consistent results can be achieved by carefully following the manufacturer's directions. To apply the developing solution, hold the ceramic object over a tray or a bucket and use a cup to flood the surface continuously for the prescribed time. Tiles and other small objects can be immersed in the solution. At the end of the recommended time, wash the surface with water or stop bath. Then paint the ceramic pigment, previously mixed, onto the moist surface of the emulsion; bathe the object in fixer; and wash it under running water until all excess pigment or silver is removed.

Images of various colors can be produced with some of the silver-based emulsions. Kallitype yields purplish brown, sepia, brown, and black tones. KPR, KOR, and Picceramic use ceramic pigment, and the resulting color will depend upon the pigment used.

Note: Because several of the photosensitizing chemicals and emulsions are toxic, well ventilated working areas are highly recommended.

Firing

Emulsions that use a ceramic pigment (oxide, glaze, or stain) should be fired to fuse the image permanently to the clay or glaze surface. A fast firing cycle gives better results with an electric kiln. For china paints and overglazes the temperature range is C/018 to C/012. Some stains and glazes can be fired to C/4. Preheating to remove any water in the emulsion and clay body prevents bubbles and pinholing.

Most fired images are transparent and can be used as an underglaze or overglaze. Several different images can be used, providing each is fired before the next is used. Other overglazes and lusters of the same maturing temperature can be fired simultaneously for interesting and unusual effects.

Protective Finish

Non–ceramic pigment emulsions like Kallitype and Vandyke use silver to create the image. Since the image is not fired onto the surface, it is subject to scratches and abuse like any paper-based photograph. The finished image can be colored with photographic toners or artist's acrylics. A protective coating of acrylic, varnish, lacquer, or a similar clear finish should be sprayed or brushed on.

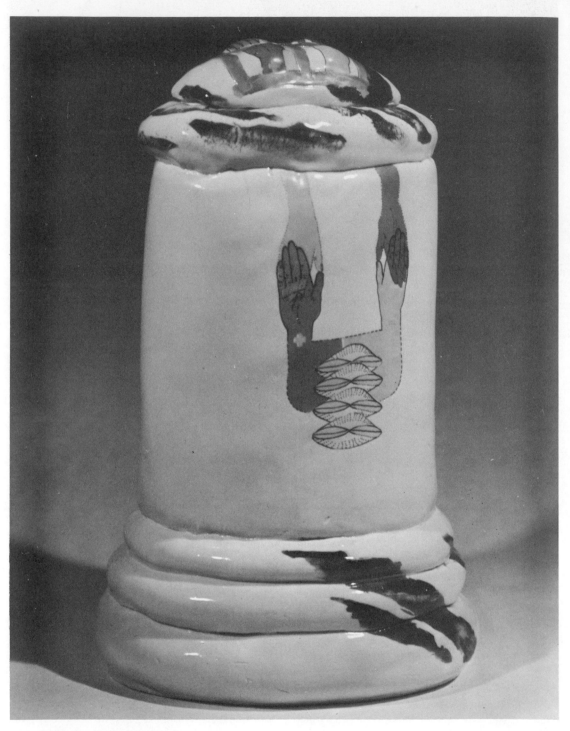

"Wonder Window," Jacquelyn Rice;
porcelain, porcelain glaze, lusters, decals; 15-1/2" H.

seven

ceramic silkscreen printing

The technique of stenciling designs on clay was known to almost all civilizations. Even in prehistoric times, we know that early potters used the technique of placing leaves on a ceramic form, applying a different color of clay over the leaves and the clay form, and then removing the leaves, revealing a leaf design. The ancient Egyptians stenciled outlines for geometric patterns onto ceramic forms and colored the patterns like a number painting.

Although stencils are adequate for many clay decorations and glazing applications, their use is limited by the necessity of holding in place the island parts of the design. The Japanese used thin strands of hair to hold the various parts of a stencil together, thus eliminating the gaps that are part of most stencil designs. Modern silkscreening, a technique in which the stencil is attached to a thin fabric, was invented in the early 1900s. The silkscreen method of printing is an extension of stenciling: the stencil is attached or supported by fiber or metal cloth stretched across a frame. The stencil blocks out parts of the cloth and leaves other parts open. A squeegee is used to force ink across the stencil and through its open areas, thus depositing ink on the printing surface. The technique is versatile in design possibilities and application. Although the ceramics industry has used this decorating technique extensively, studio potters have not.

The silkscreen process involves simple, inexpensive equipment that is easily made. The basic equipment is listed below and described later in detail.[1]

SILKSCREEN EQUIPMENT AND MATERIALS

1. Wood or metal frame
2. Screen fabric such as silk, nylon, or even organdy
3. Squeegee
4. Ceramic ink (ceramic pigments mixed

[1] This chapter covers basic equipment, materials, and techniques. For a description of other technical details and equipment, see the books listed in the bibliography.

with squeegee oil, or commercial ceramic inks)
5. Printing base
6. Stencil (paper, stencil, or photographic film)
7. Exposure lamp (photoflood, carbon-arc, sunlight, ultraviolet)
8. Miscellaneous materials: turpentine, rags, knife, masking tape, staples, gummed tape, adhering liquid, lacquer thinner, and so on.

Since the 1930s the printing of ceramic pigments on ceramic surfaces has been developed extensively. Because ceramic inks require thicker application than regular printing inks, the silkscreen process is the logical one to use. Silkscreening is the most versatile of all printing methods. Silkscreen can print delicate, sharp details, including photographs. The many processes and procedures include paper stencil, film stencil, glue stencil, photographic silkscreening, and decals and their variations. The diverse range of ceramic inks permit printing of underglaze, overglaze, glaze, stain, decal, stained glass, glass, and enamel.

There are three basic categories of silkscreening: direct, indirect, and decals. In *direct* silkscreening the screen is placed on the clay surface of green, wet, bisque, or glazeware. The design can be printed on the surface of tile, plaques, plates, glass, enameled ware, flat sides of containers, hotplates, or any other flat or slightly curved surface. *Indirect* silkscreening is a varation of direct silkscreening in which the design is printed on a thin slab of moist clay, which is then applied to the scored and slip-applied surface of bowls, dishes, containers, sculpture, or any other non-flat surface unsuitable for direct screenprinting. *Decals* are made by silkscreening the design onto a paper and then sliding the design off the paper onto the surface of bisque or glazeware. The decal can be cut, overlapped, or arranged in various designs to create unusual finished decorations not possible with direct or indirect silkscreening. Decals are used to decorate handles, spouts, lids, necks of bottles, recessed areas, and other clay, glass, or enameled surfaces.

Silkscreened designs are fired and permanently bonded to the clay or glaze surface; they will not wash off. There are many possible uses for this process, and only a relatively small number have been explored.

Equipment

A major attraction of the silkscreen process is that it does not require an expensive, cumbersome printing press; only inexpensive equipment and materials are needed. No particular carpentry skills are required to build the frame and equipment. The size of the screen depends upon its use; several screens can be used for screening small tiles, sculpture, and even large pots. The screen, the squeegee, and a sturdy table are the basic equipment.

FRAME MATERIALS

1. Dry, straight, knot-free wood strips, 2″ by 2″
2. Angle braces or corrugated fasteners
3. Flat-head $1\frac{1}{4}$″ screws
4. Heavy-duty staple gun and staples
5. Heavy-duty gummed paper tape 2″ to 3″ wide
6. Shellac
7. Scissors
8. Right-angle square

SCREEN FABRIC

The fabric generally used for the silkscreen process is silk, but metal screen, nylon, organdy, and other fibers can be used. Silk has the advantages of high stencil strength, uniform weave, and availability in mesh counts of 74, 86, 109, 124, 139, 157, and 200 openings per square inch. The mesh size used depends upon the ceramic pigment, glaze, slip, or stain used, as well as upon the type of stencil. Halftones and other fine details require a fine-mesh fabric in the 109 to 157 mesh range to give sharp edge definition. For paper stencil or direct-to-wet-clay printing, the more open-mesh fabric in the 74 to 109 mesh range is desirable because it allows the heavy glaze, engobe, or stain particles to pass through the screen. The 124-mesh screen is most useful for most screening.

Silk is sold in widths of 40 and 60 inches. The following numerical system is used to indicate the mesh count:

Number	Mesh Count
6XX	74
8XX	86
10XX	109
12XX	124
14XX	139
16XX	157
25XX	200

PRINTING BASE

A flat board is desirable for holding the screen in place. It can be made of masonite, plywood, particle board, or any similar smooth, sanded surface. Hinges hold the frame to the base; removable pin-type hinges are recommended. When extensive printing is to be done, such as decals and tile, other refinements can help speed the work. Among these are counterbalance weights, a propping arm, and quick-release hinges. The base would be a hindrance in directly applying printing to sculpture, the sides of pots, or other three-dimensional surfaces; therefore it is not used in these cases.

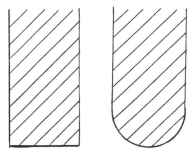

Figure 7.1

Squeegee profiles: square edge is recommended for general printing of decals, stains, or china paints; rounded edge is recommended for heavy-deposit printing of slips, glazes, or enamels.

SQUEEGEE

The squeegee is a rubber-bladed tool, one inch less in width than the inside width of the frame, that is used to draw ink across the silkscreen stencil. It distributes the ink over the screen surface, forcing the ink through the open areas of the design and onto the surface to be printed. Commercial squeegees have a rounded wood handle and a blade made of rubber that is resistant to printing inks and thinners. The rubber edge can be either square or rounded (fig. 7.1), depending upon its intended use. Even a piece of sturdy cardboard or a window-cleaning squeegee can be used for short-run printings.

Screen Construction

Ready-made screened frames are available from any commercial screen supply house. The kits and finished frames are not expensive and come in a variety of sizes. If you prefer to make your own screen, the procedure described is simple to follow.

The size of the frame should be determined before construction is started. Too small a screen will drastically limit the size of the printing area for possible later printings. A comfortable, average size is 16 by 20 inches. This is small enough for the beginner to control, yet versatile enough to allow for free printing directly on pots and for production of tiles or decals. A larger frame can be constructed to handle larger printing surfaces such as lamp bases, planters, or table tops.

Figure 7.2 is a diagram of a frame for screen printing. The frame is made of knot-free, kiln-dried wood. The four corners are constructed using the butt-end, mitered, or ship-lap techniques. To hold the corners securely and squarely together, use corrugated fasteners, screws, angle braces, or other methods. A carpenter's square will help. Loose-pin hinges or quick-release hinges are preferable for holding the frame to the base; they enable you to use several different frames with the same base, to print several colors. The base should be at least two inches longer and wider than the frame.

Many materials can be used for the printing fiber, including Swiss bolting cloth, organdy, bronze wire cloth, dacron, nylon, stainless steel, silk, and taffeta. Silk is used most frequently because it holds up best under most printing conditions. Cut a piece of fabric two inches larger than the outside dimensions of the frame. Dampen it and align it so that the warp and weft parallel the frame, to ensure proper stretch.

To staple the fabric to the frame, use a heavy-duty stapler. First drive several staples to attach the fabric to the middle portion of the bottom side of the frame (fig. 7.3). Then pull the fabric across to the opposite side and staple it to the middle portion of the frame. Repeat the procedure for the other two sides. Staple the four corners last. (This is exactly the same procedure painters use to stretch a canvas.) Do not hammer the staples all the way in until you have completely finished stapling.

After you hammer in the staples, cut away the excess fabric. Cover the inside wood area and one inch of the fabric on the bottom side with heavy-duty gummed paper (fig. 7.4). Turn the frame over and cover the inside of the frame and one inch of fabric with gummed tape (fig. 7.5). When the gummed tape has dried, paint two coats of shellac on the tape and on 3/8-inch of the edge of the fabric. The shellac and tape will act as a protective shield

Figure 7.2

Top and cross section views of a finished screen:
(A) loose-pin hinge; (B) corrugated fastener;
(C) gummed paper tape; (D) wooden frame;
(E) fabric; (F) base.

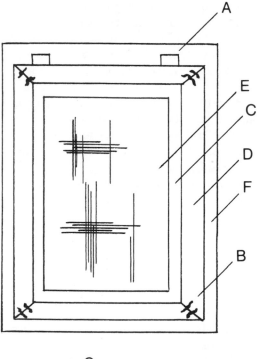

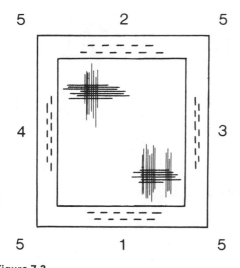

Figure 7.3

Stretching and stapling the fabric.

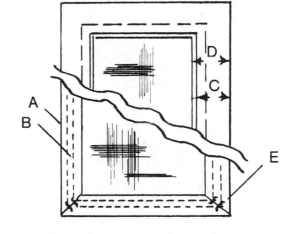

Figure 7.4

Completing the silk stretching;
taping and shellacking the underside of the frame:
(A) excess silk trimmed off;
(B) 2 alternating rows of staples;
(C) gummed paper tape covering wooden frame
and part of fabric;
(D) 2 coats of shellac; (E) corrugated fasteners.

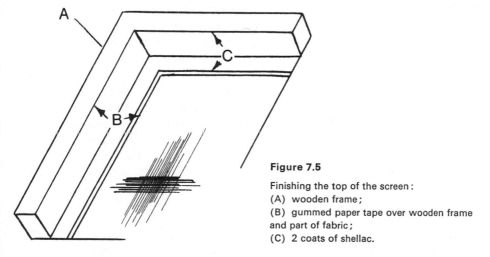

Figure 7.5

Finishing the top of the screen:
(A) wooden frame;
(B) gummed paper tape over wooden frame
and part of fabric;
(C) 2 coats of shellac.

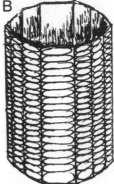

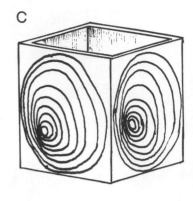

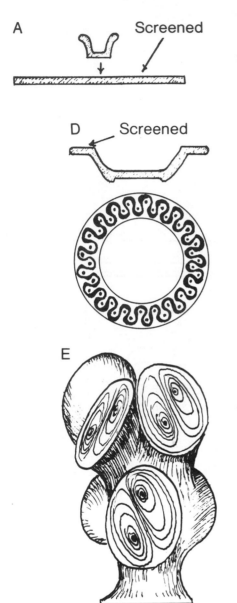

to prevent paint from seeping through the frame, and as a gathering place for excess ink. Now attach the finished frame to the base.

Optional components of the screen are a frame register guide, a leg prop, a counterbalance, and a register lock. It is not always necessary to attach the frame to the base; for example, printing of the design can be done directly onto the side of a vase, a sculpture, the lid of a jar, the rim of a plate, or the flat surface of a casserole (fig. 7.6). The surface does not have to be flat; a slightly curved surface can be screened by tilting the screen with the curve as you pass the squeegee from one area to another.

Figure 7.7 illustrates several ways of printing on a thin slab of wet clay. The slab can be left flat, domed slightly, or even stretched extensively before it is attached directly to the surface of the pot. Mix oil-based silkscreen extender with ceramic stains to prepare the medium for printing directly on soft clay. The medium will stretch with the clay. Being oil-based, it will not smudge when handled. Should any slip get on the design, it can easily be washed off. Another approach is to print on the clay slabs that are used in slab construction. These are handled just as in clay construction. Be careful, however, for the design will scratch, smudge, or come off if the slab is handled improperly.

Figure 7.6

Examples of screening directly onto finished clay form:
(A) after relatively flat areas of a lid
are screened, handle is attached;
(B) paddled surfaces of vases, cups, jars, bowls;
(C) flat sides of planters
and other slab-constructed forms;
(D) rim of plates, platters, bowls;
(E) flat or near-flat surfaces of sculpture.

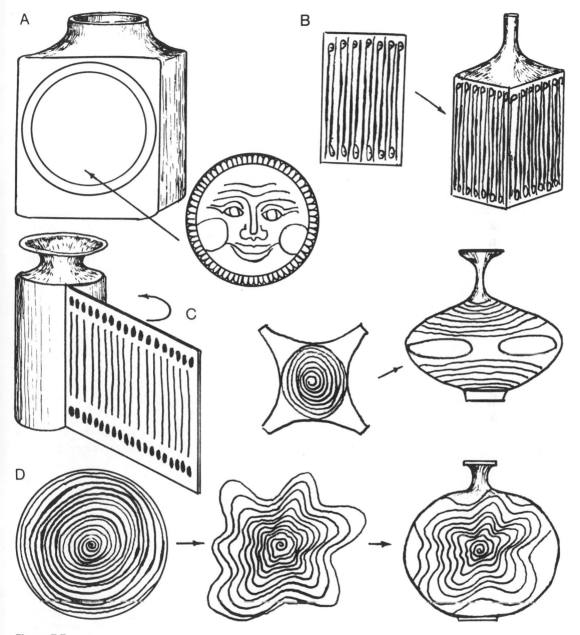

Figure 7.7

Methods of printing directly onto wet clay slab using oil-based ceramic ink:
(A) design screen-printed onto thin clay slab, then attached to vase;
(B) 4 screen-printed slabs attached to sides of vase;
(C) 2 silkscreened slabs formed into a pot;
(D) design screened onto wet clay, stretched,
and attached to clay form.

Ceramic Ink

Many silkscreen inks are used for printing on various surfaces, but only a ceramic-based ink can be used for permanent images on a ceramic surface that is to be fired. Basically, ceramic inks are a ceramic compound suspended in a liquid medium, also known as a vehicle. The ceramic compound may be stains, engobes, clays, glazes, or enamels (described below in detail). The ink is forced by the squeegee through the open parts of the design on the screen fabric, and deposited onto the ceramic form. Most commercial clays and chemicals are preground and screened and suitable for making a ceramic ink. Some commercial clay bodies contain grog, sand, or clay particles too large to squeegee through the silkscreen fabric. Some glazes, when made up, form small lumps, even though all the components of the glaze were originally fine-screened by the manufacturer. Therefore it is necessary to hand-grind or even ball-mill the compound so that it will pass through the screen mesh. The final prepared ceramic ink should be able to pass through a 200-mesh screen.

The uses of silkscreening on ceramics are extensive, depending upon the ceramic ink and the form the design is to be screened onto. Direct screening can be done on wet greenware, leather-hard greenware, dry greenware, bisqueware, glazeware, and high-fired unglazed ware. The temperature range for ceramic inks is C/022 to C/22.

Most printing on ceramics is done with overglaze or china paint ceramic inks, designed primarily for application over previously glazed clay, enamels, or even glass. These are fired at lower temperatures than underglazes or the glazed surface: between 1400° and 1700° F. Some commercial and custom-made inks are printed on top of a glazed and enameled aluminum; others are fired at even lower temperatures, such as enamels, stained glass, and glass tableware.

Underglaze is a coat of glaze, stain, engobe, or oxide applied directly to the bare bisque or greenware surface. After printing, the ware is fired, then covered with transparent glazes. Engobes and stains will not require a glaze if the clay is fired to vitrified temperatures. Most underglazing compounds are fired at clay maturing temperatures, which are generally higher than those used for the overglaze.

VEHICLES

The *vehicle* is the liquid that holds the ceramic compound in suspension so that the entire mixture can be squeegeed onto a decal or ceramic surface. Commercial ceramic inks are ready to use straight from the container. If you use custom-made inks, you must hand-mix the vehicle and the ceramic compound, using either a spatula on a glass plate or a mortar and pestle. A ball mill is used for large batches. Whether you do it by hand or with a ball mill, grinding and mixing are important. Thorough blending, wetting the ceramic compound and thickening the mixture, makes a better printing ink.

Various liquids can be used as a vehicle (medium). Any vehicle should have the following desirable qualities: (1) it should be viscous; (2) it should be safe to handle; (3) it should dry without crawling, smudging, bubbling, or developing pinholes; (4) it should not clog the screen or leave any carbon residue when fired. Commercial squeegee vehicles are highly recommended. Other vehicles work satisfactorily, but not quite as well. Several commercial and individually made vehicles are listed below.

1. *Commercial silkscreen oil-based extender.* Transparent, the best all-around vehicle that will stretch with the wet clay for

reshaping after printing. It will not smudge with handling after it is dry.

2. *Water-soluble silkscreen extender.* Its major advantage is that it will clean up with water. It forms a tough film when dry, but is not as satisfactory as the oil base.

3. *Commercial squeegee oil.* Various manufacturers make oils that have good working qualities.

4. *Decal lacquer.* Best used on lacquer-proof films. It dries fast, forming a hard film.

5. *CMC (carboxymethylcellulose).* The powder, which is mixed with water, gells overnight. A thicker mixture will take longer. The cost per quart of mixed vehicle is not high, but the cost per pound of dry powder is about $4.00.

6. *Screening varnish.* A mixture of one-third each of varnish, turpentine, and boiled linseed oil. It forms a tough film.

The following household items will work in a pinch, but they are not recommended for an extensive printing project because they will clog the screen and form a film on the ink while printing is in progress.

7. *Baby oil.* Thin, but will print satisfactorily for a short print run.

8. *Liquid soap* (such as Ivory). After about four printings, the screen starts to clog and needs cleaning.

9. *Vaseline.* This pasty petroleum product is good for short print runs.

10. *Wax emulsion.* A liquid wax that will clog the screen after a few printings.

CERAMIC INK COMPOUNDS

Silkscreen printing on ceramic surfaces is similar to silkscreen printing on other surfaces, but the inks are very specialized. Even the several commercial inks require uniform experimentation and recording of all data to develop a palette of colors and surfaces.

Engobe is a clay slip coating on a white or colored ceramic surface. A glaze is sometimes applied over the engobe, or it may be fired as is. Engobes normally are not difficult to use because they have a wide firing range and are easily applied. The difficulty is that silkscreening deposits such a thin engobe coating that it may not survive if the kiln atmosphere destroys it or the clay body and glaze coating absorb it. The engobe should adhere to the clay body and not flake off during drying and firing, and it should vitrify at the same or a slightly lower temperature than the clay body. These two considerations are minor compared to the problem of thickness. You can overcome this major problem by (1) using a rounded squeegee blade to allow more engobe to flow onto the clay body; (2) using a thicker paper stencil; or (3) adjusting the engobe composition by adding more colorants, opacifiers, or fluxes.

A preferred engobe is composed of the particular clay you are using, with the addition of colorants. A slip of the clay is strained to remove sand, grog, and lumps. Then the colorants are added. The screened slip is set to dry, broken up, strained again, and finally mixed with the vehicle. This engobe must be applied only to wet or damp clay, for it may loosen and peel off when it shrinks if it is applied to a dry or bisque-fired clay. It is difficult to change a dark clay body to a lighter value without drastically changing the clay body. Additional colors can be made by adding opacifiers and colorants.

There are many commercial and custom-made engobe preparations.[2] To make the engobe printing ink, weigh the engobe and colorant (the same colorants that are used in

---

[2] For engobe formulas, see Glenn Nelson, *Ceramics, A Potter's Handbook* (New York: Holt, Rinehart and Winston, 1971); Daniel Rhodes, *Clay and Glazes for the Potter* (Philadelphia: Chilton, 1970); and John Conrad, *Ceramic Formulas: The Complete Compendium* (New York: Macmillan, 1973).

glazes); dry mix and screen through a 200-mesh screen; and add vehicle until the mixture is thoroughly blended to a paste consistency.

Stains are certain metallic oxides and spinels. *Colorants* are oxides, sulfates, dioxides, and carbonates of metals used to give color to ceramics; for example, iron, copper, cobalt, and nickel. *Spinels* in ceramics are synthesizer minerals made by firing a mixture of the constituent oxides at sufficient temperatures to calcinate them into a crystal resistant to the actions of the minerals in glazes. Stains can be used for coloring underglazes, overglazes, glazes, clays, and decals.

The only practical way to achieve some colors like maroon red, vanadium yellow, pink, and Chinese red is to use commercial stains because of the extensive procedure necessary to custom-make these colors. The procedure consists of weighing the chemicals; firing them in a crucible to 2300° F.; removing the hot crucible from the kiln and dumping the compound into water; wet-grinding it until it will pass through a 60-mesh screen, washing it with water, and continuing to grind it until it will pass through a 200-mesh screen; and then drying it. However, some stains are easy to custom-make. For example, black is a mixture of 25 percent each of iron, copper, cobalt, and manganese oxides plus a small amount of glue. Brown, an even simpler stain, is made by mixing red iron oxide with glue. Both black and brown are strong stains, but if applied too thickly they will blister or flake off.

The following formulas will make simple stains suitable for overglaze, underglaze, decal, and clay body decorations.[3]

[3] For additional formulas and information on stains, see Rhodes, *Clay and Glazes for the Potter*; F. H. Norton, *Elements of Ceramics* (Reading, Mass.: Addison-Wesley, 1952); Eugene Ryshkewitch, *Oxide Ceramics* (New York: Academic Press, 1960); and Conrad, *Ceramic Formulas*.

Black

Ceramic black	70
Frit	30
	100

Chrome Green

Chrome oxide green	40
Zircopax	30
Frit	30
	100

Green

Copper carbonate	50
Tin oxide	25
Frit	25
	100

Purple-Brown

Manganese carbonate	40
Zircopax	30
Frit	30
	100

Dark Blue

Cobalt carbonate	55
Frit	30
Tin oxide	15
	100

Tan

Rutile	60
Tin oxide	20
Frit	20
	100

Rust Brown

Iron oxide red	60
Zircopax	15
Colemanite	15
Albany slip	10
	100

White

Tin oxide	40
Calcined kaolin	30
Frit	30
	100

The use of stains in all temperature ranges will produce hundreds of colors. Using both commercial and custom-formulated stains offers versatility and allows you to try a variety of ceramic silkscreen techniques. Stains are strong and should be screened thin; if they are applied too thickly, the colors will bleed or the protective glaze may crawl off. The ingredients in the overglaze influence the resulting color. For example, zinc retards pink colors; lead produces different colors than do alkaline-based glazes; the atmospheric character of the kiln changes some copper-based stains from blue to red; and glazes that mature at different temperatures have different color influences. The color of a stain can be lightened by using more vehicle or increasing the amount of opacifier. When the screened design is placed on clay and the ceramic ink contains an oil base, the oil will repel the overglaze. For this reason the screened design is often placed on unfired clay and then bisquefired to burn off the medium.

Glazes are thin coats of a glasslike compound that fuse to the surface of clay during firing. Glazes range from opaque to transparent; colored to colorless; gloss to matt; flat, even color to speckled; light-reflecting luster to matt-cratered; and low-temperature brilliant hues to the subtle off-whites of porcelains. To custom-formulate glazes you need a large inventory of clays, minerals, oxides, and frits. The oxides of metals and nonmetals are subdivided into bases, neutrals, and acids. A chemistry background is helpful in calculating chemical analysis empirically to determine the formulas of ceramics, since this is a specialized, technical phase of the art.[4]

Frit and raw glazes have a wide temperature range, from C/022 overglaze to C/22 porcelain, and an extensive range of types including opalescent, alligator, crater, speckled, crackle, matt, gloss, luster, clear, and opaque. There are two basic categories of glazes: frit glaze and raw glaze. A *frit* is a glasslike compound containing most of the ingredients that make up a glaze. It is melted to over 2000°F. in a smelter. The molten glass is poured into water, shattered into fine particles, then placed into a ball-mill and ground to fine powder that will pass through a 200-mesh screen. Few frits are used by themselves as glazes, for they do not contain all the necessary clay. Frits usually do not adhere to the clay body unless a binder is added. A typical frit glaze contains frit(s), clay, and a binder such as bentonite or gum. *Raw* glaze generally does not contain frits, but contains all the minerals, clays, fluxes, and binders necessary to make a glaze. Raw glazes must also be ground in a ball-mill and screened.

Using glaze to silkscreen is simple. However, when some glazes are applied too thin they are absorbed by the clay, the underglaze, or the atmosphere of the kiln during firing, to the point that the glaze may not be distinguishable. Also, fine details of a design may be lost when glaze is applied to a clay body containing a heavy amount of metallic oxides. For these reasons the screen used for applying glazes should be coarse, to permit a thicker glaze coating. Slip glazes can be applied directly to greenware, and regular glazes can be applied to bisqueware or previously glazed ware. Some designs can be printed up as decals, but the thickness of the glaze deposit makes transfers difficult.

[4] For information on how to calculate glazes and to use precalculated glazes, see Herbert H. Sanders, *Glazes for Special Effects* (New York: Watson-Guptill, 1974); Steven Goldberg, *Glaze Calculation* (San Jose, Ca.: Billiken Press, 1972); Harry Fraser, *Glazes for the Craft Potter* (New York: Watson-Guptill, 1974); Rhodes, *Clay and Glazes for the Potter*; and Conrad, *Ceramic Formulas*.

Prepared overglaze and china paints are usually vivid, low-temperature glazes that fuse to fired glaze surfaces and bisqueware. They are commonly used to decorate chinaware, often over a white or light-colored background. The overglaze consists of china paints mixed with a vehicle to a pasty consistency. Use a flexible spatula on a hard surface to blend the materials until creamy. Thin coats of overglaze produce light tints; thick coats are translucent to opaque. (Make sure the glaze is thick enough to cover the surface!) China paints make excellent decals for bisqueware or glazeware. Most of these glazes are fired slowly to C/019 to C/010.

Prepared ceramic screen inks, available in many colors, can be printed like most other screen inks. Colors such as bright metal lusters, mother-of-pearl, and red are produced at low temperatures starting at 850°F., while porcelain temperatures begin at 2300°. Most prepared ceramic inks are designed primarily for application over previously glazed surfaces, enamel, and glass. These inks are fired at lower temperatures than underglazes and other glazes. The ink is ready to use straight from the container, as long as the ink materials have not separated; before use, stir the contents thoroughly. To thin, add more oil medium, but *not* turpentine.

Prepared ceramic screen inks are ground with an oil-based medium and used directly for printing by photo screen, film screen, decal, and paper stencil techniques. The mesh number and the pressure of the squeegee determine the thickness of the ink, which should be applied thinly and evenly. When the ink dries, in about 30 minutes, additional printing can be done over the first coat. Excess application may cause blistering or bubbling.

Prepared enamels are available in dry and moist form. The dry form should pass through a 200-mesh screen. Mixed with a printing medium to the consistency of thick syrup, it should run off a palette knife in a slow, continuous stream. If it has been mixed thoroughly, the mixture will not need to be passed though the screen. Moist prepared enamels are like prepared ceramic screen inks, ready to use straight from the container.

The enamels are not much different from overglazes and prepared ceramic screen inks. They are designed for temperature ranges of 900° to 1600°F., and for use on glass, enameled metal, glazes, enameled aluminum, and clay. Since enamel is dense it holds its edge definition for multicolor designs, making accurate registration possible. Light-colored and white backgrounds are used. Used as an underglaze, some enamels interfere with over-enamels by modifying or overpowering the colors. Enamels and overglazes can be used as an underglaze for different effects. Two different transparent enamel colors can be mixed to produce a third color, but when opaque enamels are mixed, the result will be speckled like pepper.

Prepared liquid and dry glazes can be used to glaze ceramics like a printing ink. If the liquid glaze is not moist enough to use as ink, add a small amount of water-soluble medium; if it is too wet, pour the liquid into a pan and set it in the sun to thicken. Add a small amount of medium to dry glaze and mix with a palette knife on a hard surface. Most glazes can be screened, but some loss of color or glaze quality may occur because of the thinness of the glaze deposit. Some glazes have a high percentage of ingredients that burn out during firing, causing the loss of fine design detail. For a thicker coating of glaze, use a more open mesh and a rounded-edge squeegee.

Wax resist, a different silkscreen printing approach, involves the use of wax emulsion instead of a ceramic ink to print a design. The wax is screened onto the surface of greenware, bisqueware, or glazeware; then a glaze, stain, underglaze, or slip is painted, poured, or sprayed over the entire surface. The wax design resists the thin coating, and the surface under the wax shows through clearly. Using thick glaze or slips will build up small surface areas on the wax to create a beaded effect.

Stencil

There are many stencil methods, including paper stencil, cut-film stencil, tusche and glue, photographic film, glue and lacquer, water-soluble fillers, direct emulsion, masking tape stencil, rubber cement, water-soluble film stencil, and combinations or variations. Essentially, a *stencil* is blocked-out areas on a screen. Four basic stencil methods used in ceramics are paper, cut-film, direct glue, and photographic stencil.

PAPER STENCIL

To use the paper stencil method, place a paper design against the underside of a silkscreen, then apply ink to the top and squeegee. The ink holds the paper in place and prints the design onto the ceramic surface. The materials you will need are a clean silkscreen, squeegee, masking tape or gummed tape, knife or razor blade, scissors, paper, and ceramic ink (fig. 7.8).

Most paper is suitable for use as a stencil; however, unbonded typing paper is the most successful because it is strong and thin, cuts easily, and slightly absorbs the ink. Take a piece of paper several inches larger than the planned design; cut the design with a sharp knife, scissors, or a paper punch, or tear it by

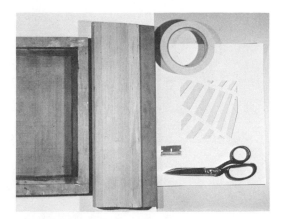

Figure 7.8

Materials needed for paper stencil technique.

hand. Avoid making folds and wrinkles. Center the finished paper design under the silkscreen, position loose design pieces; and on top of the screen block off the areas around the design with paper and tape (fig. 7.9). The cut or open areas will print.

Figure 7.9

Silkscreen showing the paper design held in place with masking tape.

Figure 7.10

Use one even pull of the squeegee to avoid smudging.

The image can be printed directly on wet or dry greenware, bisqueware, glazeware, and decals. Put several dabs of ceramic ink on the screen and squeegee, causing the paper to attach to the screen. Whenever possible, use one continuous pull of the squeegee to avoid moving the paper stencil and smudging the printing image (fig. 7.10).

Various ceramic inks can be used, including glazes, stains, and engobes. Be careful when you apply ceramic ink to ceramic ware, because the designs are easily smudged. Likewise, underglaze and stains, when fired at low temperatures (under C/04), will smudge. To prevent smudging use a transparent gloss or semi-gloss glaze to protect the design (fig. 7.11).

Figure 7.11

Tile screened with paper stencil; clear overglaze.

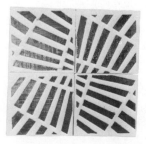

The advantages of the paper stencil include the following:

Inexpensive materials

Fast preparation

Little equipment necessary

Easy application of stencil to screen

Any ceramic ink suitable

Fast clean-up

The disadvantages of the paper stencil are the following:

Short print life (5 to 30 prints)

Not suitable for complicated designs

Tendency of paper design to move on screen if handled carelessly

Not effective if screen dries out

To clean the screen, place it on a pad of newspapers; remove the excess ink and save it in a glass jar with a tight lid, for later use. Remove the stencil, block-out tape, and paper from the screen. The stencil cannot be used again. Clean the screen with laundered rags and proper solvents, depending upon the ceramic ink. Use several newspapers and fresh rags until the screen is thoroughly clean. Whenever you can, print the entire paper stencil print run at one time.

CUT-FILM STENCIL

Precise edges, thin lines, large color areas, and free-flowing or geometric designs are characteristics of the cut-film stencil. Film manufacturers make Inko, Nufilm, Ulano, and Craftint brands, some of which are water-soluble or lacquer-proof. The cut-film method is the most common one used for making posters, serigraphs, and other commercial silkscreen prints. The finished cut-film stencil is similar to paper stencil except that the film is made of lacquer, gel, or other materials on a waxed paper or plastic base. The materials needed for cut-film stencil printing are film, a clean screen, squeegee, a sharp blade, adhering liquid, gummed tape, cleaning solvent, cleaning rags, newspaper for clean-up, ceramic ink, and the ceramic surface to be printed (fig. 7.12).

Cut a piece of film two inches longer and wider than the original stencil design; place the film over the design; and tape it down at the corners. Remove the areas to be printed by cutting an outline with a sharp-pointed blade; use just enough pressure to cut through the film. (Too much pressure will cut completely through the base sheet, making the film difficult to adhere.) When you have finished cutting, carefully peel away from the backing sheet the design areas to be printed. Keep all loose pieces of film off the surface.

Position the finished stencil face (film) up under the screen, centered to allow ample room to use the squeegee. *Note*: All silk, new or old, must be clean and dry. Soak a soft, clean rag in the adhering liquid, used to fuse the lacquer film to the screen. Rub the rag on an approximately four-inch square in the middle of the film; then dry immediately with a clean rag. Continue this procedure to the edges of the film until the entire surface is adhered. *Warning:* Do not allow too much liquid to stand on any part of the film, for it will burn (dissolve) the film. When you finish, wait about 15 minutes and carefully peel off the backing. The screen is now ready to be masked (fig. 7.13).

Inko Aquafilm is a water-soluble cut-stencil film for use with oil, lacquer, vinyl, or any ceramic ink that does not contain water. The film is cut like lacquer-based film, but the adhering process is slightly different. Dampen the clean screen with water; place the cut-film under the screen on a hard surface; soak a clean rag in water and dampen the center four-inch square. Blot the area quickly with a dry rag. Dampen and dry the balance of the film until the film is adhered.

Figure 7.12

Cut-film stencil printing materials.

Figure 7.13

Design adhered to screen (above) and masked with gummed tape (below).

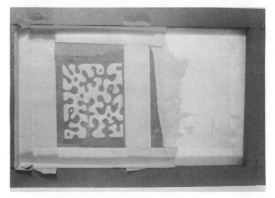

Then allow the film to dry and peel off the backing. A fan will speed drying.

Mask the open areas around the adhered film with a liquid masking agent, gummed paper tape, or masking tape and paper (fig. 7.13). Apply ceramic ink to the top of the screen, on the masked area. The heavy paste ink will not drip through the screen, but the pressure of the squeegee will force the ink through the mesh onto the ceramic surface. Hold the squeegee at an angle between 50 and 80 degrees, and apply firm, even pressure (fig. 7.14). If possible, pull the squeegee across only once, because the screen may shift slightly and the second pull may blur the image.

Overprinting, multicolors, repeats, repositions, and other uses of the same design can produce many design variations (fig. 7.15). The advantages of the cut-film stencil technique include the following:

More complex designs possible than with paper stencils

Produces sharpest, cleanest printed edges of all the silkscreen techniques

Print run of several hundred possible

Supplies simple and few

Figure 7.14

Pull the squeegee firmly, once across the silkscreen.

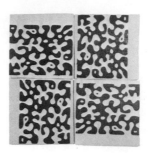

Figure 7.15

Variations in cut-film stencil screened design on tile; clear overglaze.

Screen with stencil attached can be cleaned and stored for later use

Reasonable cost

The disadvantages of using a cut-film stencil are the following:

Danger of dissolving the film while adhering it to the screen

Halftone or shading not possible

Most inks dry in a short time; engobes, which are thicker, take longer to dry—longer still if they are applied to moist or wet greenware. Remove surplus ink from the screen and store it in a container for later use. Lay the screen over several flat sheets of newspaper. Pour a small amount of solvent directly onto the stencil to remove the ceramic ink; rub the screen with fresh, soft rags until it is clean. The clean stencil can be stored and re-used several times.

To remove the film stencil from the screen so that the screen can be used with another stencil, put several layers of newspaper on a table. Remove tape and block-out paper from the screen. Saturate the underside of the stencil with the proper solvent and immediately place it on top of the newspapers. Apply additional solvent to the surface of the screen and rub with a rag to saturate and soften the film. Then lift the screen; the film will stick to the newspaper. Place the screen on fresh newspaper and use additional solvent to remove any remaining film. Now the clean screen is ready for re-use.

Cut-film. Amber, NuFilm, Pro-Film, and Inko 340, 500, 520, and 540 brands are all-purpose films with waterproof backing. All of them are removed easily from the screen, handle fine details, adhere with stencil-adhering liquid, and can be removed with lacquer thinner. All are available in local art supply shops.

Aquafilm. Inko 547 brand is water-soluble stencil film that can be knife-cut and used with any ceramic ink that does not contain water. Warm water adheres the film; solvent-adhering liquids are unnecessary. Hot water is used for clean-up.

DIRECT GLUE

The direct glue technique is well suited to free drawing, retaining the freshness and spontaneity of brushwork. The materials for this technique (fig. 7.16) are a clean silkscreen, squeegee, brush, mucilage, ceramic ink, block-out materials, and the ceramic surface. Place a clean screen over the original design and brush the background areas not to be printed with full-strength mucilage. When

Figure 7.16

Materials for direct glue printing.

Figure 7.17

Full-strength glue applied directly to the screen.

the screen is partially dry, hold it to a light to check for pinholes. Fill any pinholes and dry completely (fig. 7.17). (If you make a mistake, wash the entire screen or parts of it with water, dry it, and start over.) Block out the areas around the design with paper and tape, or cut a paper mask and position it under the screen (fig. 7.18).

Figure 7.18

Background area blocked out and paper mask positioned under the screen.

Figure 7.19

Screened design on 2 wet greenware tiles;
tile on right was reshaped and both were fired
with clear overglaze.

Applying glue with a dry brush, an atomizer, a toothbrush, an airbrush, or texturing tools produces broken effects of stippling, spattering, shading, and other textures. These textures are among the most desirable aspects of the direct glue technique. The glue can be force-dried with a fan; however, avoid high heat. The image can be printed directly on wet or dry greenware, bisqueware, glazeware, or decals (fig. 7.19). Any ceramic ink except water-based ink is suitable. Apply the ink to the screen and squeegee to attach the paper mask (if used) to the screen (fig. 7.18).

The advantages of the direct glue technique include the following:

Fast preparation

Little equipment, few supplies

Inexpensive materials

Application of glue to screen very simple

Unusual textures possible

Fast clean-up

The disadvantages of the technique are these:

Glue breaks down on long print runs

Water-based ink unsuitable

Difficult to produce sharp edges in design

To clean the screen, remove the block-out paper and tape. Use the proper solvent to remove excess ink from the screen, and set it aside to dry for future use. To clean the screen completely, place the screen in a sink and flush the surface with hot water while scrubbing gently with a soft bristle brush.

Photo Silkscreen

With a photographic stencil the ceramist can go beyond the scope of the knife-cut, paper, film, and direct glue stencils to produce fine line and photographic detail. Both halftone and color separations can be produced with accuracy. Several of the photographic silkscreen processes are relatively simple and not overly expensive, can be produced quickly, and will withstand the use of abrasive ceramic inks. The ceramist can have a local printer make the halftone photographic stencil (or transparency), or he can make it himself.

Photographic stencils can be made several ways. The manufacturers of the materials supply detailed instructions on the use of their products. There are carbon tissue, presensitized film, direct screen emulsion, and transfer photo stencils; each type has advantages and disadvantages. Most types of photo stencils are exposed through a high-contrast positive transparency, which is the reverse of the photographic negative. In a high-contrast positive transparency the opaque areas are free of thin spots or pinholes. Transparent areas are clean and not smudged, and halftone areas have separate, clean dots.

PHOTO POSITIVE

The most important aspect of photo silkscreening is the positive used to create the image on the sensitized photo surface. There are many approaches to making the positive, from the straightforward technique of painting opaque ink on transparent plastic to the halftoning of the negative using a copy camera. A description of the methods of producing the photo positive follows.

Professional printer A professional printer makes a halftone or line positive from a photographic negative or from the original copy. Look in the yellow pages of the telephone directory or in the list of suppliers at the back of this book for the names of printers. Using a printer is the easiest way to produce halftones and other positives; however, the cost of a finished 8- by 10-inch positive can be as much as $10.00.

Ink positive To make an ink positive, place a piece of transparent acetate on top of your original design. Reproduce the design on the transparency by applying opaque ink with a brush or a pen, and allow the ink to dry. (On transparent bases such as Mylar or Estar you can use 4B to 8B pencils, grease pencils, or other opaque media.) The clear areas of the transparency expose the stencil film below, leaving the unexposed areas to print the image. Figure 7.20 shows the ink positive and the materials for making it.

Figure 7.20

Ink positive (right); original design (left) and materials for producing positive.

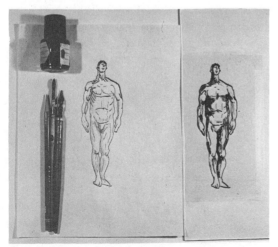

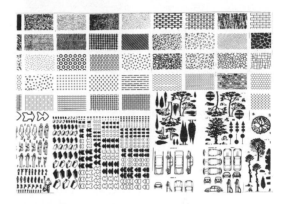

Figure 7.21

Commercial image and press-type transfers.

Transfer positive Press-type, lift-type, and design transfers; commercial opaque letters; and any other opaque designs can be glued, pressed, or applied in any other way to the transparent plastic base (fig. 7.21).

Acrylic lift positive Making an acrylic lift positive by lifting ink from a magazine photograph is easy and inexpensive. The materials you will need are acetate, denatured alcohol, acrylic spray, warm water, a brayer, and a magazine photograph. Most high-quality magazines print halftone photographs on flint- or kaolin-coated paper. Choose a photograph or design from such a magazine and blot the surface with denatured alcohol to loosen the ink. Spray an acetate (plastic)

Figure 7.22

Acrylic lift positive (right)
made from magazine photograph (left).

sheet with clear acrylic spray. When the acrylic becomes tacky, place it on the photograph, using slight pressure with a brayer. After two hours or so, when the paper is dry, soak the acetate in warm water. Rub the back of the magazine page gently until the paper comes away, leaving the ink on the acetate. The positive is now ready to be used (fig. 7.22). Some dots are so small that it is difficult

to produce a clear silkscreen print from them. It may be necessary to use a halftone screen on the film when you print the photo silkscreen film from such an acrylic lift positive.

Thermofax positive   To make a thermofax positive, position your original design on top of Agfa Transparex film and run it through a thermofax copier (fig. 7.23). A positive transparency will emerge. Two or more thermofaxed sheets sandwiched together may be required to achieve solid blacks. Only those images, photographs, ink drawings, and magazine photographs printed on one side of the paper may be used. (This is because light is transmitted through the images and any other image on the reverse side will also expose the thermofaxed sheets.) The complete photo silkscreen process using the thermofax positive method is presented in figures 7.25 through 7.32.

Photo line positive   The photo line positive process translates a photographed image into solid black or white areas. Because the film that is used is not sensitive to grays, it renders them as solid black or white, depending upon their density. Commercial photo labs and printers convert negatives or original designs into line positives; however, you can do it yourself in a darkroom. You will need an enlarger and Kodalith Ortho #3 film in addition to the usual darkroom equipment. The original design can be a painting, a

Figure 7.23

Using a thermofax copier to make a thermofax positive:
(A) Transparex film; (B) opaque image, facing up;
(C) paper holding opaque image.

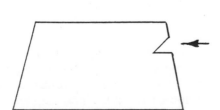
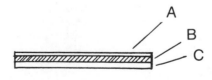

drawing, a photograph, or any other image that will produce a photographic negative. The larger the negative size, the higher the quality of the finished silkscreened image. With a copy camera or enlarger, the negative is enlarged onto high-contrast film, then processed using developer, stop, and fixing chemicals. Depending upon the enlarger and the negative quality, small-sized film negatives (35mm, for example), when they are projected to make a transparency larger than 8 by 10 inches, may lose some of their detail and sharp edges.

High-contrast film is available in 8- by 10-, 11- by 14-, 16- by 20-, and 40-inch rolls from Kodak, DuPont, GAF, and 3M. The base of the film is acetate or Estar (Kodak); Estar is recommended for large sizes because of its ability to hold its shape during processing.

Halftone positive The halftone positive method renders gray values by breaking up the photographic areas into black and white dots of varying sizes. Producing a halftone positive is the only way to obtain a gray scale. When a design such as a painting or photograph contains gray areas ranging from light to dark shades, the halftone positive method is necessary to prepare the design for the silkscreen process. The multitude of dots determines the gradual tonal change of light to dark in the halftone. Ranging from large dots close together for the dark shades, to medium dots for middle shades, to small dots spaced more widely for the light shades, the dots gradually change in size and spacing, reflecting the original design.

A halftone screen is a transparent sheet containing opaque horizontal and vertical lines in different equally spaced patterns, creating a finely ruled screen separating the image into thousands of squares ranging from a low of 40 to a high of 300 lines to the inch. Higher-numbered screens are too fine for ceramic inks; the recommended range is 60 to 100 lines to the inch. *Note:* The screens,

which are expensive, should be handled with care; one scratch will ruin the surface.

Several methods are available to produce the halftone:

1. *Contact halftone copy.* The halftone screen is positioned close to the film in the copy camera, and the design is photographed. The image on the film, made up of halftone dots, is known as the halftone negative. The negative is used to make a transparent positive.
2. *Halftoning photographic positive film.* The halftone screen is placed in front of the object to be photographed or in front of the film in the camera. The film produces a positive rather than a negative. Available in 35 mm size, the positive is used to make black-and-white 35 mm slides; it requires processing slightly different from negative film processing. The processed transparent positive is projected onto the photo stencil film using a slide projector or an enlarger.
3. *Halftone photographic film.* This type of film, which has a halftone screen directly on it, will produce the halftone dot pattern without the use of a separate halftone screen. The halftone negative transparency is projected onto a Kodalith film, making a halftone positive transparency that is ready for use.

Full-color designs similar to color magazine photographs and lithographs are possible through the use of color filters that separate the various colors of the design into their basic magenta, yellow, cyan, and black ink components. Each color requires a separate halftone stencil. Commercial decals are made using the full-color technique. (See the section on ceramic decals, later in this chapter.)

DIRECT PHOTO EMULSION METHODS

Any subject or material can be photographed to make the negative from which a photographic stencil is produced. For that reason this technique has an aesthetic as well as a versatility appeal for ceramists who want to create unusual, personalized surfaces not possible with any other ceramic technique. The technique is appropriate whether the ceramist's orientation is toward the "pop" movement, which uses images directly from cartoons, comics, and other mass communication media, or toward the subtlety of a fine line drawing. The innovative possibilities are limitless.

Several approaches can be used for photo silkscreening, using either direct or indirect emulsion methods. The direct emulsion is the least expensive method, but it requires several steps to prepare the screen. A disadvantage of direct emulsion is that the emulsion sensitizer must be used the same day as the emulsion is mixed and the sensitized screen must be exposed within a few hours after the sensitizer is applied to the screen. The indirect method requires a separate film and costs slightly more than the direct method but there are fewer limitations imposed by the materials.

The direct emulsion method, a good all-purpose photographic stencil method, works on silk, nylon, and other mesh screens. Metal or synthetic fabric screens are recommended; silk is not, because the bleach used to clean the screen weakens silk fibers. A major advantage of the direct emulsion method is that the emulsion resists all ceramic inks including varnish, lacquer, oil, and water-based ink. Also, fine details and long production runs are possible because the emulsion is embedded in the screen mesh rather than adhered to the bottom.

The equipment necessary for the direct emulsion process ranges from the basic screen to the vacuum table, copy camera, photo enlarger, and motorized screen printers, depending upon the complexity of the project. An opaque projector can be used to project the original design onto a screen sensitized with Rockland Emulsion SC-12.

MATERIALS AND EQUIPMENT FOR DIRECT PHOTO EMULSION METHODS

1. Silkscreen, cleaned of all grease, lint, dust, and old ink by scrubbing with mild abrasive cleaner (Ajax) or trisodium phosphate. Be sure to rinse away the cleaner with hot water. Commercial washes (Azokleen or Mesh Prep) are effective and efficient.
2. Direct screen emulsion: Enko #71, Azocol R, Dichromate, Rockland Superfast Emulsion SC-12.
3. Squeegee.
4. Photo enlarger (optional).
5. Photo equipment: tray, measuring cup, containers, sponge, fan.
6. Light source: fluorescent, photoflood, carbon-arc, ultraviolet, sunlight, or photo enlarger.
7. Emulsion remover: household bleach or Azokleen.
8. Positive transparency: halftone or line positive.

Preparing the screen. To prepare the screen, use a sensitizing solution according to the manufacturer's directions. Mix the two components until thoroughly blended to make up the emulsion. Avoid shaking, which causes air bubbles. If you add a colored tint, you will be able to see the emulsion more clearly while squeegeeing it onto the screen. Rockland SC-12, made with silver nitrate, must be handled under amber safelight darkroom conditions. Other sensitized emulsions can be mixed and used under ordinary dim incandescent electric light.

Coating the screen. Pour the prepared emulsion onto the edge of the screen, between the tape and the silk. Using a squeegee that fits the screen, pull with a light, even stroke to distribute the emulsion. Clean the excess off the edges and the squeegee. Fan dry. Apply a second coat to the underside of the screen, then clean the excess off the edges and the squeegee. The extra emulsion can be used as a block-out. Air or fan dry, then check to determine any pinholes. (Pinholes can be filled by using a toothpick dipped into the mix.) Unused emulsion can be stored for later use. When the screen is completely dry, make the exposure promptly.

Exposure. The exposure is made with a photo enlarger or a light source: sunlight, fluorescent, photoflood, ultraviolet lamp, or carbon-arc. Carbon-arc requires the shortest exposure time. Exposure times range from two to ten minutes. Good contact is essential between the glass, the transparent positive, and the screen; a vacuum table is desirable, but a weight-and-glass system will suffice. To make the contact print exposure, sandwich the glass, the positive, the emulsion-coated screen, a felt blanket, and a board, with weights to hold everything in place. The sandwich can be placed face down on an

exposure table or reversed to face an overhead light source (fig. 7.24). If you use a photo enlarger, place the positive in the enlarger and project it onto a sandwich arrangement consisting of a glass, the emulsion-coated screen, a felt blanket, and a board, with weights to hold all in place (fig. 7.24).

Developing procedure. Use cold or warm water (100° F.) to wash exposed areas under subdued light. Cover both sides of the screen with water to soften the emulsion. A soft spray of water will remove most of the softened emulsion; scrub any remaining emulsion gently with a nylon brush. Blot excess water with newsprint and allow the screen to dry. Touch up with screen block-out.

Emulsion SC-12 is more difficult to use than the other emulsions. It requires a photo developer, and it must be washed in 110° water and stop-bathed. Wash the screen to remove any excess, blot it with newsprint, and allow the screen to dry.

Figure 7.24

Cross section of contact print arrangement (left)
and photo enlarger print arrangement (right)
for direct sensitized silkscreen:
(A) light source; (B) weights; (C) glass;
(D) transparent positive; (E) sensitized silkscreen;
(F) table; (G) felt blanket; (H) board; (I) photo enlarger.

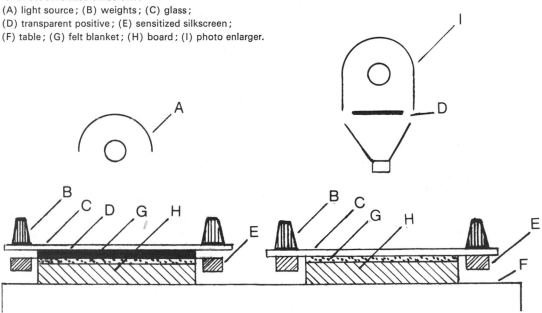

Cleaning the screen. Use mild bleach, hot water, lacquer thinner, or hydrogen peroxide, depending upon the emulsion. Be careful, because bleach and hydrogen peroxide weaken silk fibers. Wear rubber gloves.

INDIRECT EMULSION METHODS

The indirect emulsion stencil is the best means of easily and accurately producing silkscreen prints with fine details of line copy, halftone, and color separation. The films used in this technique are presensitized on a plastic backing. They require little treatment before exposure and remove easily from the screen.

Many manufacturers sell presensitized film for both contact and photo enlarger exposures that will adhere to silk, nylon, and other mesh screens. All ceramic inks except water-based inks can be used. These films are dimensionally stable and easy to use, and they stand up to long production runs. Hi-Fi Green and Prep 220 films do not require a darkroom.

The materials and equipment needed are as few as a screen, photo trays, and developers. Photo enlarger and darkroom conditions are necessary with some of the films and processing techniques.

MATERIALS AND EQUIPMENT FOR PRESENSITIZED PHOTO FILM METHODS

1. Clean silkscreen
2. Presensitized photo film: Ulano Hi-Fi Green, Carbon Tissue (requires sensitizing), Craftint 200, DuPont 260, McGraw 4272, Prep 220, Rubylith
3. Film developers (DuPont 260 uses activator, rinse, and stop bath chemicals)
4. Photo lab equipment: developer tray, measuring cup, storage containers
5. Squeegee
6. Ceramic ink
7. Light source: fluorescent, photoflood, carbon-arc, ultraviolet, sunlight, or photo enlarger
8. Transparent positive: halftone or line

Photo-stencil films. These are presensitized films that produce fine details. Several brands are available. Ulano Hi-Fi Green, Craftint 200, Prep 220, and Rubylith do not require special darkroom treatment; Carbon Tissue needs to be sensitized with potassium dichromate. DuPont 260 is a presensitized camera-speed photographic film on a plastic base, requiring photographic development and stop bath treatment under darkroom conditions with a ruby lamp. The other films are handled and processed under low incandescent light.

Developers. The developers for some films are two-part mixes in powder form; other developers are pre-mixed in a single solution. DuPont 260 requires developer, activator, and stop bath solutions. The compounds are weighed out (some are available in premeasured packets) and added to premeasured amounts of water, mixed to dissolve the crystals, and poured into a tray large enough to hold the film. Use just enough liquid to cover the film.

Exposure. Contact printing follows almost the same exposure process as the one used for the direct emulsion method. A sandwich of a glass plate, a transparent positive, a sensitized photo-stencil film, a felt pad, and a board, with weights holding it all in place, is set face down on an exposure or vacuum table or reversed to face the overhead light source (fig. 7.25). If you use the photo enlarger, place the positive in the enlarger and project it onto a sandwich of glass plate, sensitized photo-stencil film, felt pad, and board, with weights to hold all in place (fig. 7.25). Again, a vacuum or exposure table can be used to hold the sandwich

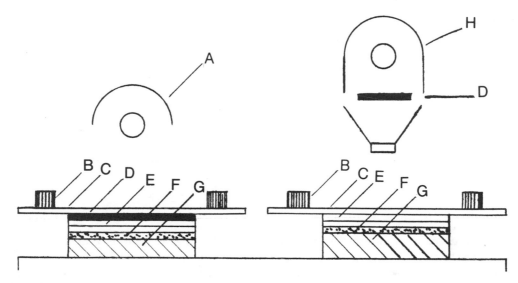

Figure 7.25

Cross section of contact exposure (left)
and photo enlarger exposure (right)
of photo-stencil film:
(A) light source; (B) weights; (C) glass plate;
(D) transparent positive;
(E) sensitized photo-stencil film; (F) felt pad;
(G) board; (H) photo enlarger.

Figure 7.26

Arrangement for contact exposure:
glass plate, thermofaxed transparent positive,
sensitized stencil film.

arrangement. Good contact is essential
between the glass, the transparent positive,
and the film. Figure 7.26 shows the proper
arrangement of glass, transparent positive,
and film.

Light source and time tables. The exposure time for sensitized films ranges from 10
seconds (in an enlarger using DuPont 260
film) to 60 minutes (using Craftint 200 under
a photoflood lamp). Various sources of
actinic light can be used, including carbon-
arc, photoflood, ultraviolet sunlamp, yellow
exposing bulb, fluorescent lamp, photo
enlarger, and sunlight. Certain substances
in the emulsion placed on acetate backing
become resistant to water when sensitized
with bichromate and exposed to *actinic light*
(light of shorter wave length, such as violet
and ultraviolet light). The energy of the

Don't miss the 'CISCO KID' on Television every Tuesday 7-7:30 P M.

Figure 7.27

Guide to light source, distance, and exposure time

Light Source	Distance from Light to Film	Time of Exposure
Ultraviolet sunlamp (250-watt)	30 inches	14 minutes
Yellow exposing bulb (for DuPont 260)	30 inches	10 seconds
Carbon-arc lamp (30-amp)	30 inches	2.5 minutes
Photoflood #2	30 inches	15 minutes
Sunlight	93,004,000 miles	3–8 minutes
Black light fluorescent lamps (bank of 8)	4 inches	3 minutes

actinic light produces a chemical change in areas exposed to light; parts of the film not exposed to light are washed out with water.

The sandwich faces the light source. Make a test strip to determine the proper exposure time, which depends upon the type of light and the distance from light to film. To avoid heat build-up on the film, you may have to use a fan. Halftone exposures must be precisely timed because of their fine dot structure. The longer the exposure, the thicker the film will become. If it is exposed too long the film will lose fine detail. Figure 7.27 is a guide to distance and time of exposure for various light sources. Different films require different exposure times. Manufacturers' directions give exact details.

Developing procedure. Developing is done under subdued incandescent or amber light. Wear rubber gloves to protect your hands. Pour into a photo tray just enough developer to cover the film. Gently rock the tray every 15 seconds for one to two minutes, depending upon the manufacturer's directions. If you keep it covered, you can use the developer for several films and discard it at the end of the day. (Some brands of developer can be stored in a cool, dark place, but they gradually lose their strength.)

Wash the developed film with a gentle spray of warm water (110° F.) until the design is clear of all unexposed gelatin. Finish by rinsing the film with chilled water.

Adhering the film. Use a clean screen for proper adhesion. Place the wet, washed film emulsion-side up on a smooth, flat, hard surface. Wet the silk fibers slightly with a sponge, carefully position the screen over the film, and gently lower it to make contact with the film. Press lightly by hand with clean newsprint. Then use a roller to blot thoroughly, removing excess water and smoothing out the film (fig. 7.28). Use additional newsprint until the film is adhered to the screen. Avoid too much pressure! A fan speeds drying. When the film is completely dry, carefully peel the backing sheet from it. Fill any pinholes or make corrections with a liquid block-out, and mask out the open areas around the film with gummed tape or liquid block-out (fig. 7.29).

Printing. Use any ceramic ink that does not contain water. Screen it through a 200-mesh screen onto greenware, bisqueware, or glazeware (figs. 7.30 and 7.31). Figure 7.32 shows the clay slab, reshaped and attached to a ceramic container and fired.

Figure 7.28

Use a roller to blot and smooth the wet film to the screen.

Figure 7.29

Mask areas around the dry, adhered film with tape.

Figure 7.30

Image inked onto moist greenware.

Figure 7.31

Original design (left); thermofaxed image (middle);
screened image on wet clay (right).

Figure 7.32

Fired container, silkscreen photo-film technique.
"Cisco Kid," Ron Carlson; stoneware, oxide rub,
and screened image; 17" H × 15-1/2" W.

Cleaning the screen. It is much easier to clean up from indirect emulsion printing than to remove emulsion from the screen fabric. Remove the excess ink with a piece of cardboard and put the ink in a container. Clean the screen thoroughly with clean rags wetted with the proper ceramic ink solvent. After the screen dries it can be re-used. To remove the film from the screen, scrub with a nylon brush and flood with hot water. Cleaners like Inko SPS, Ajax, or other, similar mild abrasives will remove any stubborn emulsion. A final wash with vinegar will neutralize the cleaners.

Figure 7.33

German beer steins, circa 1860–1910.

Ceramic Decals

Decalcomanias or printed transfers, called decals, are designs or photographs printed on specially prepared paper from which the design may be transferred to surfaces of clay, glass, or enamel. When more than one copy of a drawing, design, or photograph is desired, the silkscreen printing process is an obvious advantage. The decal can be placed on ceramic surfaces onto which direct screening cannot print because of surface quality or texture or irregular shape of the clay form. The decal can also be placed in a recessed area that cannot be printed upon directly.

An indirect printing method used as early as the 1740s, the art of decalcomania reached its zenith in France and England in the latter part of the nineteenth century. Figure 7.33 shows German beer steins decorated with decals, circa 1860–1910.

Ceramic decals, unlike ordinary decals, are made with ceramic ink. Stains, china paint, and glazes can also be used. Decals are manufactured commercially (fig. 7.34) for decorating dinnerware, enamels, glass, tile, electronic circuits, artware, and stained glass.

Figure 7.34

Commercial decals.

Figure 7.35

Custom decals.

Custom decals (fig. 7.35) have several advantages and produce effects not available from commercial decals. With the silkscreen photo-stencil process you can reproduce drawings, designs, found objects, photographs, photo-collage, fabric, paintings, and comics to make decals instead of drawing, painting, or decorating directly on the clay form.

The equipment and materials used for making decals are the same as those used for cut-film, glue, and photo-stencils. The only addition is the decal paper.

BASIC EQUIPMENT AND MATERIALS FOR DECAL METHOD

1. Decal paper: Simplex, Duplex
2. Squeegee
3. Ceramic clear-coat varnish
4. Ceramic ink: underglaze, stains, or overglaze mixed with silkscreen printing media
5. Clean silkscreen
6. Turpentine, lacquer thinner, rags
7. Prepared film and its adherent materials
8. Block-out, masking paper, or gummed paper tape

DECAL PAPER

Commercial decal paper (brands such as Simplex and Duplex) is water-absorbent paper coated with a water-soluble adhesive gum. Although commercial paper is better, you can make decal paper by coating a smooth, water-absorbent paper with a thin layer of starch.

STENCIL

Almost any silkscreen stencil process can be used to make a decal, but the photographic halftone and line techniques produce the most attractive effects. Decals produced by halftone and line techniques enable ceramists to transfer photographs, delicate drawings, and unusual designs to most ceramic surfaces, whereas silkscreening is limited to relatively flat surfaces.

For full details on the silkscreen process, see the earlier sections of this chapter. What follows is a brief outline of the photo-stencil technique of preparing the screen for making decals (see figure 7.36). The original design,

Figure 7.36

From original image to finished decal:
original image (top left);
negative enlarged with copy camera (middle);
transparent positive (top right);
image inked on decal paper (bottom left);
varnish-coated decal, ready for use (bottom right).

drawing, photograph, painting, or other subject matter is photographed to make a negative. This can be done with a regular camera or with a copy camera if an enlarged negative is desired. The negative is used to make a transparent positive by the technique of contact negative (fig. 7.25), a projected negative using a photo enlarger (also fig. 7.25), or a negative with Agfa Transparex film in a thermofax machine (fig. 7.23). Either the transparent positive is projected onto the photo-stencil film using a photo enlarger, or a full-sized positive of the desired image size is used to make a contact print on the film. The contact print is sandwiched in an arrangement of glass, positive, and photo-stencil film, facing the source of actinic light. After exposure the film is placed in the developer for the required time, then washed under running water. The unhardened areas are washed away from the backing. A clean screen is positioned over the wet film, which is placed on a hard surface. Newsprint absorbs the water and a brayer smooths, blots, and lightly pressures the film onto the screen (fig. 7.37). When the film is dry, the backing is gently pulled from the film (fig. 7.38). The open areas around the film are then masked (fig. 7.39), leaving the screen ready to print decals.

PROCEDURE

By printing your own decals you can create designs and subject matter not available commercially, thus increasing the potential use of the process. Making all the decals at once saves a great deal of time; once the screen and materials are set up, the actual printing goes very fast.

1. Check the adhered film for pinholes and defects. Correct them with block-out liquid.
2. Mark the position for decal paper under the screen on the printing base with three

Figure 7.37

Film is blotted and smoothed to screen using a brayer and newsprint.

Figure 7.38

Backing is peeled away from film.

Figure 7.39

Areas around film are masked.

Figure 7.40

Image is inked to decal paper.

Figure 7.41

A coat of varnish is applied over
the inked image.

Figure 7.42

Cross section of finished decal:
(A) varnish coat; (B) inked image;
(C) soluble gum; (D) stencil paper.

registration tabs, to ensure the exact location
for subsequent repetitive printings.

3. Cut decal paper to the desired size. Place
   one piece under the screen.
4. Some ceramic inks are sticky and may hold
   the decal paper to the screen. Use double-
   faced masking tape or a loop of masking
   tape, or spray adhesive onto the board under
   the decal paper, to help hold the paper
   in place when the screen is lifted.
5. Apply the premixed ceramic ink along
   the top edge of the stencil and squeegee it
   over the image with one or two quick, even
   strokes (fig. 7.40). If the ink thickens after
   several prints, add fresh ink to thin the mix.
6. Lift the screen and let the printed decal
   dry for 24 hours. The decal can be reprinted
   several times for multicolor designs.
7. When the decal is dry, brush or screen
   a coat of ceramic clear-coat varnish over it.
   Cut a paper stencil one-quarter inch larger
   than the design, to squeegee the varnish over
   the ink design (fig. 7.41). Depending upon
   the ceramic ink you use, you may have
   to screen a piece of newsprint between each
   varnish printing to remove the ghost image
   left by the previous decal printing.
   Figure 7.42 shows a cross section of a
   finished decal.
8. Allow the varnish to dry thoroughly and
   store the decals between pieces of wax paper.
9. After printing the decals, clean the screen
   with the proper solvent to remove the ink.
   Put the screen away until later, or clean it
   with the proper solvent to remove the film
   permanently from the screen.

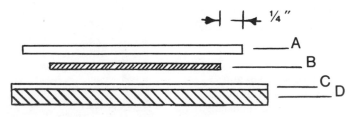

DECAL TRANSFER

Cut the decal from the sheet and carefully trim it of excess paper. Take care not to mar the pigment area. Immerse the decal in warm water for 30 seconds or until the image is loose. Position the decal for application on the ceramic form, and hold one side of the decal in place; now slide the backing paper from under it. The image remains on the surface of the ceramic form. Remove air bubbles and wrinkles with a soft cloth or roller, starting in the center and pressing outward. Blot excess water with facial tissue. Be careful—repositioning is difficult, and any portion of the decal not firmly attached to the surface will burn away during firing. Although overworking the decal will cause tearing or stretching, you can turn this effect to your advantage by producing a controlled distortion of a photo image.

FIRING

After a day or so, when the decal has dried *completely*, the ceramic form is fired (fig.7.43). Any moisture under the decal will curl or pop it. Fire the kiln at low temperatures with its lid or door propped open three inches for ventilation for the first two hours, to allow the fumes from varnish and ceramic inks to escape. After two hours, close the door and continue firing slowly to avoid blistering, to allow combustion of organic matter, and to produce proper fusion of the decal. The firing temperature depends upon the ceramic ink you are using.

COMPLEX DESIGNS

More than one decal can be positioned on the same ceramic form, or the decal can be used as an underglaze. Other glazes, stains, or lusters can be applied around or over decals and fired with the decal firing. The

Figure 7.43
Fired decal on glazed tile.

decal and other glaze must have the same fusing temperature. When a decal is used as an underglaze, the surface of the clay must be smooth and non-porous. Raku and low-temperature clays will absorb the water from the decal rapidly, making transfer of the decal difficult; first seal the surface with a varnish or decoupage glue.

Silkscreened Transfers

Les Lawrence, of El Cajon, California, has refined the process of silkscreened transfers. His technique is to screen an image onto tissue paper and immediately press the tissue gently onto moist clay, transferring the image from the tissue to the clay.

Any technique for screening an image will work well. Start with a screen and an image prepared for printing and ceramic ink. Many oil-based ceramic inks using stains, china paints, or glazes are successful. Lawrence recommends an ink made of Pemco #812 black stain and Craftint Crystal Clear vehicle plus a few drops of kerosene, mixed to a screening consistency with a palette knife on a glass plate. Screen the image onto tissue paper and immediately position the tissue on the wet to leather-hard clay, because the ink dries quickly. Gently rub the tissue with a soft rubber roller or the back of a spoon until the image transfers. Treat the transferred image on the clay surface as if it were a glaze or stain. Glazes can be applied over the transfer; several transfers can be overlapped on the clay; other stains and glazes can be applied and fired; and transfers can be placed over glazes.

The screened transfer technique has many advantages. It is a quicker process than decals; it allows you to place images on irregular surfaces more easily than decals; an undesired image is easily removed from the clay with lacquer thinner; and you can carve designs and lines through the transferred imaged to expose the clay. There are disadvantages, too: the screened transfer image is less sharp than that produced by decals or direct screening; and only one image can be printed at a time, because the ink dries quickly on the tissue.

Transfers

Although the transfer technique is technically not part of silkscreening, it was the forerunner of the decal process. Transfers were first used in the late seventeenth century on English and French enamels. Among the very earliest production ceramics to use the transfer process was the "chinoiserie" made at Worcester (Bristol) China Works in England. This famous enterprise started making porcelain in 1748. The factory entered into direct competition with the Far Eastern porcelains then flooding the European market. Not only did the European potters follow the forms of Far Eastern ceramics, but as much as they could they adopted similar decorative themes and techniques. The Chinese used the transfer process as a major method of decorating ceramics, and so did the Europeans.

Three variations of the transfer process were used at Old Worchester. (1) An image was printed on paper; then the paper was placed on the clay and rubbed, transferring the wet ink to the surface of the ceramic form. The ceramic inks formed an outline of brown or black ceramic pigments, which were colored over with brightly colored enamels. (2) Several color separations on different plates were printed on paper, then transferred individually to the ceramic surface. The resulting image was similar to a color lithograph. (3) A design was printed on paper using an oil-based medium, then transferred to the surface of the ceramic form by rubbing the paper onto the surface. Enamel was dusted over the wet oil and the excess was blown away. Each enamel color was fired separately.

PROCEDURE

Any relatively hard surface with a relief pattern design, such as a wood cut, linoleum cut, etching plate, rubber stamp, wood grain, or leather, can be reproduced. Place silkscreen medium or squeegee oil on a glass slab and roll it out with a good brayer, using horizontal and vertical strokes to distribute a very thin layer on the brayer. With even pressure, deposit a layer of oil (medium) onto the textured surface. Place wax paper over the oiled design and rub the surface with a clean brayer. Carefully remove the paper, position it on the ceramic surface, and rub with a clean brayer. Lift off and discard the paper. Use fresh wax paper for each transfer to avoid smudged designs. Dust the oiled surface with dry ceramic pigment, then remove the excess by blowing or by brushing gently with a soft brush. The ceramic pigments can be stains, underglazes, glazes, overglazes, or enamels. Unglazed bisqued surfaces may not be suitable, since they may absorb the oil quickly. Apply a thin coating of glue to the clay to seal the surface; then apply the transfer.

"Made in U.S.A. B#1," Les Lawrence;
stoneware, underglaze, silkscreen, matt and gloss glazes; 23" diam.

"Man Feeding Birds," David Stewart;
stoneware, underslips, matt glaze; 9-1/2" H.

ceramic portfolio

"Floor Vase," Richard Peeler;
stoneware, matt glaze;
28-1/2" H × 24" diam.

"Covered Jar," Steve McGovney;
porcelain, matt blue crystalline; 13" H.

"Tan with Blue Crystals Vase," Laura Andreson;
porcelain, crystalline glaze;
10-3/4" H × 4" diam.

"Bottle," Roger D. Corsaw;
stoneware, applied clay, glazed;
10-1/2" H.

"Hanging Planters and Bowl," Janette Rothwoman;
colored stoneware, rope, natural, matt glaze;
bowl 7-1/2" H × 11" diam.

"Four-Quart Casserole, Large and Small Canisters," Harriet Cohen;
stoneware, celadon with stamped decoration;
large canister 9-1/4" H.

"The Salamander Enlightenment of Clouds in the Sky,"
Philip Cornelius; porcelain,
underglaze, clear glaze over; 12" square.

"Quasi Colombian, Almost African," William Wilhelmi;
stoneware, beads, hair, lead weights,
stoneware glaze; 15" H.

"Pyramid Double Play," Patrick McCormick;
porcelain, stoneware, natural, gloss glaze; 38 H".

"The Housewife's Dream,' Larry Elsner;
bisque-fired stoneware, smoked, burnished surface;
6-1/2'' H × 14-1/4'' W.

"Monument to Political Prisoners," Maurice Grossman;
stoneware, natural glaze, rope; 41'' H.

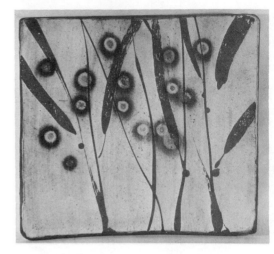

"In Wildness," David Shaner;
stoneware, engobe, natural, ash spots;
13-1/2" H.

"Lidded Jar," Dale Roush;
stoneware, matt glaze; 11-1/2" H × 6-1/2" diam.

"Wall Piece," Vivika Heino and Otto Heino;
stoneware, iron rub, matt white glaze; 16" square.

"Table Setting," Sylvia Hyman;
stoneware, matt and semi matt; plate 8-1/2'' × 10''.

"Vase Form," Jean Yates;
stoneware, oxides, gloss glaze, slips; 8-3/4" H.

"Vase," Ron Korczynski;
stoneware, white matt, iron decoration; 14-3/4" H.

"Branch Box," Michi Itami Zimmerman;
porcelain, natural, oxide; 18-1/2" H.

Untitled, Harrison McIntosh;
stoneware, matt glazes; 7'' H × 6'' diam.

"Bottle Form," Herbert H. Sanders;
porcelain, orange and gold crystal glaze; 11-1/2'' H.

"The Alhambra Tube," H. James Stewart;
stoneware, wax resist, glaze, stains; 17-3/4" H.

"Blue Jacket," Marilyn Levine;
stoneware, engobes, natural; 32-1/2" H × 19" W.

"Ceramisan," William Shinn;
stoneware, skin, wire, wood, black semi-matt glaze;
10" H × 46-1/2" W.

"Bottle," Dorothy Bearnson;
stoneware, gloss glaze, luster, gold leaf;
9-1/4" H × 8" diam.

"Seven-Piece Canister Set," Tyrone Larson;
stoneware, tan glaze and iron stain;
10" H × 18-1/2" W.

"Lollipop Head," Ernie Cabat;
stoneware, wax resist with matt stoneware over;
12" H.

"Xacnal," Claude Conover;
stoneware, engobes; 17-1/4" H × 10-3/4" square.

"Bottle," Frank Ross;
stoneware, matt glazes; 18-3/4" H.

"Bottle," Herb Schumacher;
stoneware, Aspen ash glaze; 5-1/2" H × 5-1/2" diam.

"Porcelain Pendants," Allan Widenhofer;
porcelain, underglaze, C/06 clear glaze over;
box 12-3/4" H × 16" W.

Untitled, David Krouser;
stoneware, black and dark tan matt glaze; 58" H.

"Branch Bottle," Hacik Gamityan;
porcelain, Swedish glaze; 3-1/2'' H.

"Umbilical Series," Wesley Mills;
stoneware, stain, matt glaze; 5-3/4'' H × 7'' diam.

"Covered Colander," Val Cushing;
stoneware, matt glaze; 10-1/4'' H × 11'' diam.

"Mother and Child," Raul A. Coronel;
stoneware, engobe, natural; 41'' H.
Photo by Raul A. Coronel.

"Dinnerware Setting," William J. Gordy;
stoneware, "Mountain gold"; plate 10'' W.

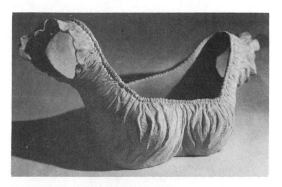

"Running Blouse," Jerry Caplan;
earthenware, white engobe, natural;
11-1/2" H × 24-3/4" W.

"Porcelain Slab Bottle," Oscar Bucher;
porcelain, matt white glaze, stain; 18" H.

"Lidded Container," Roger Bailey;
stoneware, white and saturated iron matt glaze;
17-1/2" H × 12-1/2" diam.

"Branch Bottle," Edward Cromey;
stoneware, glaze, oxides, salt-fired; 12" H.

"Goodby, Columbus," Susan Howsare;
stoneware, acrylic paint; 14" H.

"Porcelain Bowl," Charles Lakofsky;
porcelain, natural, stamp design;
6" H × 7-1/2" diam.

"Box Planter," Bob Nichols;
stoneware, gloss glazes with over-decoration; 28" H.

"Vessel Form with Stripes," George Timock;
Raku, black and white variation; 15" H × 16" W.

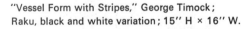

"Totemic Form #2," Ron duBois;
stoneware, yellow matt glaze; 24" H.

"Bowl," Richard Devore;
stoneware, stoneware and low-temperature glazes;
11-3/4" H × 11-3/4" diam.

"Hooker," Stephen Zawojski, Jr.;
stoneware, salt glaze, frits; 13-1/2" diam.

"Floor Urns," Donald J. Sutherland;
stoneware, matt glaze, natural, oxide;
19-1/2" to 21-1/4" H.

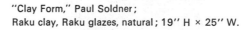

"Teapots," Byron Temple;
flameware, matt glazes; 9" H and 7" H.

"Clay Form," Paul Soldner;
Raku clay, Raku glazes, natural; 19" H × 25" W.

"Teapot," Bill Sax;
porcelain, celadon and saturated iron glaze; 11-1/2" H.

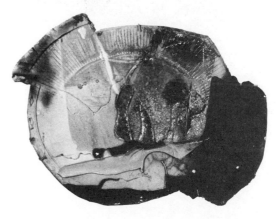

"Ecological Jig I (left) and II (right),"
John Stephenson; aluminum and porcelain, natural
and glazed; 25-1/2'' × 29-1/2'' × 1-3/4'' deep;
30'' × 22-1/4'' × 2-1/4'' deep.

"Jar," Kenneth Ferguson;
stoneware, salt glaze; 22'' H.
Photo by Kenneth Ferguson.

"Sculpture with Black and White Slip,"
Theodore Randall; stoneware, slips;
38-1/4'' H × 33'' W.

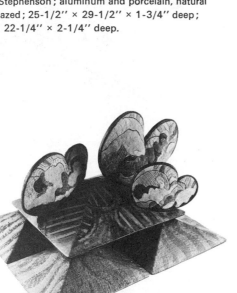

"Inscape #2," Jerry Rothman;
Ferro-ceramic, natural, glazes, stains;
32'' H × 48'' square. Photo by Jerry Rothman.

"Large Globular Vase," Fong Chow;
stoneware, orange and white glaze;
13-1/4" H × 13-1/2" diam.

"Synergism I," Robert Piepenburg;
stoneware fired Raku, Raku glaze; 19-3/4" H.

"Teapot," Angelo Garzio;
stoneware, matt glazes; 10" H. Photo by Angelo Garzio.

"Bottle," J. Sheldon Carey;
stoneware, C/9 reduction white and black iron glazes;
10" H. Photo by J. Sheldon Carey.

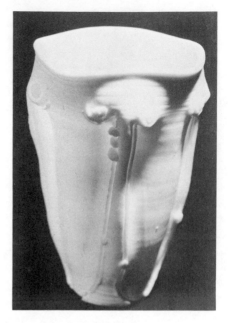

"Light Gather," Rudolf Staffel;
porcelain, natural; 7-1/2" H.

"The Vikings," Elaine Katzer;
stoneware, white matt glaze, iron oxide, natural,
pencil; 72" H.

"Never," Daniel Rhodes;
stoneware, natural; 30" H × 17-1/4" W.

"Plate," Peter Voulkos;
stoneware, natural, sprinkled glaze; 18-1/2" diam.

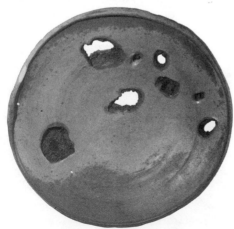

bibliography

Low-Temperature Glazes and Glaze Calculation

ANDREWS, A. I., *Ceramic Tests and Calculations* (New York: John Wiley & Sons, 1955).

BEHRENS, RICHARD, *Glaze Projects* (Columbus, O.: Professional Publications, 1972).

BINNS, CHARLES, *The Potter's Craft* (New York: D. Van Nostrand, 1947).

CONRAD, JOHN, *Ceramic Formulas: The Complete Compendium* (New York: Macmillan, 1973).

"Facts about Lead Glazes for Art Potters and Hobbyists" (New York: Lead Industries Association, 1972).

"FDA Laboratory Information Bulletin No. 834" (Washington, D.C.: U.S. Food and Drug Administration, Division of Compliance Programs, Bureau of Foods).

FRASER, HARRY, *Glazes for the Craft Potter* (New York: Watson-Guptill, 1974).

GOLDBERG, STEVEN, *Glaze Calculation* (San Jose, Ca.: Billiken Press, 1972).

GREEN, DAVID, *Understanding Pottery Glazes* (London: Faber & Faber, 1963).

LAWRENCE, W. G., *Ceramic Science for the Potter* (Philadelphia: Chilton, 1972).

LEACH, BERNARD, *A Potter's Book* (London: Faber & Faber, 1960).

"Lead Glazes for Dinnerware, ILZRO Ceramics Manual #1" (New York: International Lead Zinc Research Organization, Inc., 1970).

NELSON, GLENN, *Ceramics, A Potter's Handbook* (New York: Holt, Rinehart and Winston, Inc., 1971).

NORTON, F. H., *Elements of Ceramics* (Reading, Mass.: Addison-Wesley, 1952).

PARMELEE, CULLEN W., *Ceramic Glazes* (Chicago: Industrial Publications, 1951).

RHODES, DANIEL, *Clay and Glazes for the Potter* (Philadelphia: Chilton, 1970).

RYSHKEWITCH, EUGENE, *Oxide Ceramics* (New York: Academic Press, 1960).

Glass

SHAND, E. B., *Glass Engineering Handbook* (New York: McGraw-Hill, 1958).

WEYL, WOLDEMAR, *Coloured Glasses* (London: Dawson's of Pall Mall, 1959).

Raku

LEACH, BERNARD, *A Potter in Japan* (London: Faber & Faber, 1960).

Lee, Sherman, *Tea Taste in Japanese Art* (Greenwich, Conn.: Asia House Gallery, 1961).

Nakamura, Julia, *The Japanese Tea Ceremony* (Mount Vernon, N.Y.: Peter Pauper, 1965).

Piepenburgh, Robert, *Raku Pottery* (New York: Macmillan, 1972).

Rhodes, Daniel, *Kilns: Design, Construction, and Operation* (Philadelphia: Chilton, 1968).

Riegger, Hal, *Raku, Art and Technique* (New York: Van Nostrand Reinhold, 1970).

Sanders, Herbert H., *The World of Japanese Ceramics* (Palo Alto, Ca.: Kodansha International, 1967).

Salt-Glazing

Ball, F. Carlton, "Clay Bodies for Salt Glazing," *Ceramics Monthly*, April 1962, 13, 32.

McWhinnie, Harold, "Some Thoughts on Salt Firing and Kilns," *Pottery Quarterly*, 11, no. 42.

Parmelee, Cullen W., *Ceramic Glazes* (Chicago: Industrial Publications, 1948).

Rhodes, Daniel, *Clay and Glazes for the Potter* (Philadelphia: Chilton, 1959).

———, *Kilns: Design, Construction, and Operation* (Philadelphia: Chilton, 1968).

Salt Glaze Ceramics (New York: American Craft Council, 1972).

Webster, Donald B., *Decorated Stoneware Pottery of North America* (Rutland, Vt.: Charles E. Tuttle, 1972).

Luster

Bedford, John, *Old English Lustre Ware* (New York: Walker, 1958).

Fraser, Harry, *Glazes for the Craft Potter* (New York: Watson-Guptill, 1973).

Parmelee, Cullen W., *Ceramic Glazes* (Chicago: Industrial Publications, 1951).

Rawson, Philip, *Ceramics* (London: Oxford University Press, 1971).

Shaw, Kenneth, *Ceramic Colors and Pottery Decoration* (New York: F. A. Praeger, 1962).

Troy, Jack, "Fuming in the Salt Kiln," *Crafts Horizons*, 32, no. 3 (June 1972), 28, 71.

Photosensitized Ceramics

"An Introduction to Photofabrication Using Photosensitive Resists," Booklet P-79 (Eastman Kodak, Rochester, N.Y. 14650).

"How You Can Permanently Fire (1400–1500° F.) Your Most Precious Photos into the Glaze of a Fine China Plate" (Picceramic Co., Vestal, N.Y. 13850).

"New Horizons in Photography" (Rockland Colloid Corp., Piermont, N.Y. 10968).

"Photofabrication Methods with Kodak Photosensitive Resists," Booklet P-246 (Eastman Kodak, Rochester, N.Y. 14650).

"Photographic Sensitizer for Cloth and Paper," Booklet AJ-5 (Eastman Kodak, Rochester, N.Y. 14650).

Williams, Gerry, "Photoresist," *Studio Potter*, 1, no. 1 (Fall 1972), 12–15.

Silkscreen

Auvil, Kenneth, *Serigraphy* (Englewood Cliffs, N.J.: Prentice-Hall, 1965).

Biegeleisen, J. I., *Screen Printing* (New York: Watson-Guptill, 1971).

Carr, Francis, *A Guide to Screen Process Printing* (London: Vista Books, 1961).

Chieffo, Clifford, *Silkscreen as a Fine Art* (New York: Van Nostrand Reinhold, 1967).

Heller, Jules, *Printmaking Today* (New York: Holt, Rinehart and Winston, 1972).

Inko Screen Process Methods, Materials and Techniques (Oakland, Ca.: Screen Process Supplies Co., 1962).

Kosloff, Albert, *Ceramic Screen Printing* (Cincinnati: Signs of the Times, 1970).

———, *Photographic Screen Process Printing* (Chicago: Naz-Dar Co., 1962).

Peterdi, Gabor, *Printmaking* (New York: Macmillan, 1971).

Ross, John, and Clare Romano, *The Complete Printmaker* (New York: Free Press, 1972).

sources of supplies

Bienenfeld Industries West, Inc., 321 North Figueroa, Wilmington, Ca. 90744 (Sheet colored glass)

Blinko Glass Co., Milton, W.Va. 25541 (Sheet colored glass)

Colorado Art Glass Works, 1516 Blake St., Denver, Colo. 80202

Ferro Corp., 4150 E. 56th St., Cleveland, O. 44105

Keystone Cullet Co., 436 Willow Crossing Rd., Greensburg, Pa. 15601

Kokomo Opalescent Glass Co., P.O. Box 809, Kokomo, Ind. 46901 (Sheet)

National Display Materials, 65 N. 6th St., Brooklyn, N.Y. 11211 (Scrap glass)

Pemco Corp., 5601 Eastern Ave., Baltimore, Md. 21202

Talmach Supply, P.O. Box 2332, La Puente, Ca. (Cullet)

T. C. Thompson Enamels, 1539 Old Deerfield Rd., Highland Park, Ill. 60035

Whittlemore Durgin Glass Co., P.O. Box 2065, Hanover, Mass. 02339

Paul Wismach Glass Co., Paden City, W.Va. 26159 (Sheet)

Raku Equipment

Robert Piepenburgh, 515 E. Windamere, Royal Oak, Mich. 48073 (Tongs)

Photo Emulsions

Eastman Kodak Co., 343 State St., Rochester, N.Y. 14650 (Booklets P-246, AJ-5, P-79, and Photoresist Emulsions)

Picceramic Co., 817 Ethel Place, Vestal, N.Y. 13850 (Picceramic)

Rockland Colloid Corp., 599 River Rd., Piermont, N.Y. 10968 (Print-E-Mulsion)

Silkscreen Materials

Advance Process Supply, 400 N. Noble St., Chicago, Ill. 60622

Apex Roller Co., 1541 N. 16th St., St. Louis, Mo. 63106 (Brayers and rollers)

Atlas Silkscreen, 1733 State St., Chicago, Ill. 60647

Brittains Ltd., 26 Strawberry Hill Ave., Stamford, Conn. 06902 (Decal paper)

California Process Supply, 2836 – 10th St., Berkeley, Ca. 94710

Craftint Manufacturing Co., East 152nd St. at Collamer Ave., Cleveland, O. 44110

General Research and Supply Co., 572 Division Ave. S., Grand Rapids, Mich. 49303

Ideal Roller Co., 21 – 39th Ave., Long Island City, N.Y. 11101 (Rollers)

McGraw Colorgraph Co., 175 W. Verdugo Ave., Burbank, Ca. 91503

Naz-Dar Co. of California, 756 Gladys Ave., Los Angeles, Ca. 90021

Process Supply Co., 1987 N. Branch St., Chicago, Ill. 60622

Screen Process Supplies, 1199 East 12th St., Oakland, Ca. 94606

Serascreen Corp., 147 W. 15th St., New York, N.Y. 10011

Ulano, 210 E. 86th St., New York, N.Y. 10028

Custom Decals

Battjes Decals, 5507 – 20th St. W., Bradenton, Fla. 33507

Philadelphia Ceramic Supply, 1666 Kinsey St., Philadelphia, Pa. 19124

Custom Photo Work

Graphic Art Service, 114 E. Lenon Ave., Monrovia, Ca. 91016

J. M. Photographic, 7400 Irvine, Pittsburgh, Pa. 15207

Paragraphics, 117 Mitchell Blvd., San Rafael, Ca. 94903

Ceramic Glazes, Pigments, and Vehicles for Silkscreening

American Art Clay Co., 4717 W. 16th St., Indianapolis, Ind. 46222

Ceramic Color and Chemical Co., Box 297, New Brighton, Pa. 15066

Drakenfield, Box 519, Washington, Pa. 15301

Duncan Ceramic Products, 5673 E. Shields, Fresno, Ca. 93727

Falcon Co., P.O. Box 71, Jennerstown, Pa. 15547 (Rolling pins and decorating tools)

Ferro Corp., 60 Greenway Dr., Pittsburgh, Pa. 15204

Firehouse Ceramics, 55 Prince St., New York, N.Y. 10012

Franklin Adams, 2102 S. El Camino Real, San Clemente, Ca. 92672

Glidden Company, 1101 Madison Ave., New York, N.Y. 10028

Leslie Ceramic Supply, 1212 San Pablo Ave., Berkeley, Ca. 94706

Mason Color and Chemical Co., East Liverpool, O. 43920

Paramount Ceramics, 220 N. State, Fairmont, Minn. 56031

L. R. Reusche, 2 Lister Ave., Newark, N.J. 07105

Standard Ceramic Supply Co., P.O. Box 4435, Pittsburgh, Pa. 15202

Van Howe Ceramic Supply, 4810 Pan American Freeway N.E., Albuquerque, N.M. 87109

Way-Craft, 394 Delaware St., Imperial Beach, Ca. 92032

Commercial Decals

Ceramic Corner, P.O. Box 516, Azusa, Ca. 91702

Commercial Decal Inc., 650 S. Columbus Ave., Mt. Vernon, N.Y. 10550

Decalcraft, Bethlehem Pike, Hatfield, Pa. 19440

Hill Decal Co., R.&R., Box 1105, Houston, Tex. 77039

Joy Reid Ceramic Studio, 2016 N. Telegraph, Dearborn, Mich. 48128

index